Mind Embodied

This book is part of the Peter Lang Humanities list.
Every volume is peer reviewed and meets
the highest quality standards for content and production.

PETER LANG
New York • Bern • Berlin
Brussels • Vienna • Oxford • Warsaw

Jay Seitz

Mind Embodied

The Evolutionary Origins of Complex Cognitive Abilities in Modern Humans

PETER LANG
New York • Bern • Berlin
Brussels • Vienna • Oxford • Warsaw

Library of Congress Cataloging-in-Publication Data

Names: Seitz, Jay, author.
Title: Mind embodied: the evolutionary origins of complex
cognitive abilities in modern humans / Jay Seitz.
Description: New York: Peter Lang, 2019.
Includes bibliographical references.
Identifiers: LCCN 2018059330 | ISBN 978-1-4331-6614-3 (hardback: alk. paper)
ISBN 978-1-4331-6342-5 (ebook pdf) | ISBN 978-1-4331-6343-2 (epub)
ISBN 978-1-4331-6344-9 (mobi)
Subjects: LCSH: Cognition. | Brain. | Cognitive neuroscience.
Classification: LCC BF311 .S4325 2019 | DDC 612.8/233—dc23
LC record available at https://lccn.loc.gov/2018059330
DOI 10.3726/b15321

Bibliographic information published by **Die Deutsche Nationalbibliothek**.
Die Deutsche Nationalbibliothek lists this publication in the "Deutsche Nationalbibliografie"; detailed bibliographic data are available on the Internet at http://dnb.d-nb.de/.

I would like to thank my mom and maternal grandmother for being there for me in the early days of graduate school in New York City. I was also enormously influenced by a girlfriend at the time who had just graduated from the Julliard School of Dance and introduced me to modern dance. To this day, I'm still trying to figure out the meaning of a Pina Bausch performance at the Brooklyn Academy of Music that she invited me to, where the stage was covered in water with dancing hippos. She affected me in a way that early love always does—she was unbelievably warm and pretty with signature spiked blond hair as well as disarmingly authentic with everyone she met—and put my artistic and professional interests on a much different path than I would have ever expected. I must also acknowledge all of the incredible graduate, law, and medical students I met from Columbia, City University of New York (CUNY), New York University, and The New School for Social Research during my graduate school days. They were bristling with ideas and intellectual chatter from parts of the world far flung, and we had countless discussions in the libraries, restaurants, and downtown cafes that still define and animate the city. Then, too, there were many graduate faculty at the New School and CUNY who greatly affected me, but one that stood out probably more than any other was the late Professor Sylvia Scribner, who had just left the Laboratory for Comparative Human Cognition at Rockefeller University, after finishing field work in Africa, to join the CUNY faculty. I was mesmerized by her drive and range of interests, and although she was up in age by this time, she always appeared to me to possess more creativity than most adolescents or young adults. Professor Howard Gardner at Harvard, who I met in the summer of 1983 at a summer conference in Toronto—I had just finished reading and studying his book on the theory of multiple intelligences and his early work on metaphor and artistic development in children—has been a very significant influence on my thinking, too.

"We shall not cease from exploration. And the end of all our exploring will be to arrive where we started and know the place for the first time."

—T. S. Eliot

CONTENTS

ILLUSTRATIONS

ACKNOWLEDGMENTS

I would like to personally thank Meagan Simpson, Acquisitions Editor, and Liam McLean, Editorial Assistant, both at Peter Lang in New York, for assistance with the manuscript and all of the details that go into writing a book. Similarly, I must thank Arwen Armbrecht, Digital Communication Manager, at Peter Lang in Berlin, for assistance in making my Blogs for this book on *Medium* even better. Most of all, I would like to thank my sister, Patricia Seitz, Head of the Graduate Program in Architecture at Massachusetts College of Art and Design and Principal at Seitz Architects, Inc., for writing support on the North Shore of Massachusetts, as well as my brother, Donald Seitz, CEO of Inkubate and Assistant Director, Alumni Engagement, Princeton Entrepreneurship Council of Princeton University, for support and encouragement.

INTRODUCTION

Mind Embodied

When I first embarked on these studies on the embodied nature of thought, like most people, I was struck by the strangeness of the idea that the body could somehow be implicated in thinking—something we typically associate with the workings of the mind-brain, as it turns out, a subaltern term. But when I began observing my own and others' actions and well as the actions of other species it was obvious that thought was, almost always accompanied by some movement, action or activity. Even the daydreamer is scanning the environment when lost in thought. She is just not particularly conscious of her awareness of the external environment suggesting that daydreaming, a common human mental activity yet unique in its own right, is a special cognitive state, somewhat akin to the phenomenon of hypnosis where one part of the mind is walled off or unaware of what another part of the mind is thinking or doing. Yet this scanning of the environment in daydreaming and many other cognitive states are not done in isolation from one's own body which, after all, is the mechanism by which sensations enter and stimulate thinking or modulate its progression in the first place. Think of Stephen Hawking, the physically challenged astrophysicist, but still reliant on his senses and head and eye movements for reading and writing.[1] So, the first principle that I will articulate in what follows is that thinking always involves some movement, action or activity. This brings me to a personal story that exemplifies the second principle.

When I was quite young, I held a fascination for science fiction film and one in particular would keep me up all night. It was the story of a 9-year-old boy who witnesses a spacecraft landing in the sand dunes behind his beachside home. His parents don't believe a word of it when he tries to tell them about it but when his father finally traipses out on the sand dunes to prove his son wrong, he returns acting strangely different. The boy notices something foreign implanted in the back of his father's neck and the same fate is soon to await his mother. Finally, he too is brought forcibly back to the dunes by his parents and the three of them are swallowed up by a large fissure that opens up in the sand below them. Beneath the dunes is a series of tunnels created by the inhabitants of the spacecraft and quickly they are escorted off by tall, dark, Giacometti-like drones to meet with their master being.[2] The master being was depicted in the movie as a head inside a glass-filled dome carried around by dutiful drones.[3] Intelligence is thus portrayed as a brain-based phenomenon. Indeed, that appears, to be the modern view of human intelligence and mentation and, quite frankly, as I will attempt to demonstrate, a mystification of human and animal cognition.

It's like a piece of government propaganda that is slowly absorbed by the tissues of the mind-brain-body (MBB) until all thought ossifies and we begin to think, almost exclusively with background assumptions that color our views on a particular subject or issue. The war in Iraq (2003) is a good example, whatever your particular views, the 2016 U.S. Presidential election, which drew strikingly black and white distinctions, and so is the idea that thinking occurs solely in the brain.

But thought is not just "in the head" as if the connections we have with the external world of other people, the environment, culture, as well as the connections with the tissues of our own body are of no fundamental importance. Deep Blue, the IBM supercomputer, was really just a fancy calculator that could perform one thing quite well (play chess) not a thinking body or embodied mind that could strategize; taunt; desire; procrastinate; move chess pieces; think about another's thoughts and in this case its opponent, Gary Kasparov, the reigning world grandmaster; overhear and react to conversations in the room; fidget in its seat; be proud or sad; tap rhythmically in anticipation on a keyboard or other input device; sweat; change its mind; orient to smells, colors or forms in the room; scratch its head in thought; and scan the room for visual or conceptual ideas.[4] Therefore, the second principle is that thinking is a completely embodied activity with intelligence distributed throughout the MBB even though certain important functions are housed—more or less—in specific areas, such as the brain or body, yet networked with each other and with the internal and external environments. Thought, then, is not just in the head.

Why would anyone think otherwise? There is a long history of writing in Western thought with a set of largely unexamined background assumptions that have profoundly colored how we think about thinking as well as many other things. For example, the role of women and men in society, the nature of religious beliefs and worship of superior or transcendental forms, and the concept of race, which is itself a misnomer as there is only one human species, *Homo (H.) sapiens*, and it cannot be further divided into subspecies or races. Instead, we prefer to talk about ethnicities or cultural groups.

The neurosciences and cognitive sciences have greatly illuminated many important functions of the central nervous system (CNS), how it develops and even how it may have evolved, but almost in complete isolation from the rest of the body. The enteric nervous system or brain of the gut, and the third division of the autonomic nervous system that regulates our internal organs, has only recently been given its due, although it appears to play a central role in many psychosomatic disease states such as gastroesophageal reflux disease (GERD), irritable bowel syndrome, and indigestion or dyspepsia. The enteric nervous system is housed within the walls of the gastrointestinal tract (GIT) and is a crucial boundary that separates us from the outside world of sundry toxins and microorganisms, among other things. These antigens—the body in its own tactical maneuvers produces antibodies to these foreign antigens—are kept in often uncertain balance within the individual through the actions of the GIT mediated by the enteric nervous system. Social capital, however, or our network of relationships with others including family, friends, co-workers, and group and community affiliations, is now thought to modulate an individual's internal mental state (our brains), which influences the hormonal milieu of the body that alters our susceptibility to infection. In this case, the central nervous system is being affected not just by our bodies but by the surrounding culture and environment. New frontiers in neuroscience and cognate fields are beginning to illuminate these intimate connections between brain, body, and culture.

The third principle is that perceiving and acting—noticing and doing—are two sides of the same coin. Even though, technically, they can be neurologically dissociated in disease states, they are the two vertices of the cognitive triangle of perceiving, acting, and thinking. Even this distinction is a false analogy because in everyday experience the three are indissolubly intertwined. What we perceive and interact with we often think about consciously or unconsciously. And what we perceive in our experience we can't fail to often engage in some kind of action, movement or activity with an object, thing or person. The exception to this rule is the intrusion of basic reflexes such as pulling one's hand away from a hot stove or the carrying out of short, ballistic movements in which great speed and efficiency very briefly trumps the cognitive triangle as prepared by hundreds of millions of years of evolution.

So, thinking (1) almost always involves some movement, action or activity, (2) is an embodied activity distributed across the MBB, and (3) is one vertex of the cognitive triangle of perceiving, acting, and thinking. Gilbert Ryle, the British philosopher, believed that Cartesian dualism or René Descartes' claim that mind and body were different substances was false; mind was just the *ghost in the machine*.[5] Descartes was the 17th century French philosopher who famously said, "Ego cogito, ergo sum" or "I think, therefore I am." His argument for the belief in our own cognitive powers was grounded in the "evil demon" thesis. If an evil demon could deceive him about the information from his senses and the external world was merely a mirage, then the only thing that could sway him with surety was the reality of his own thoughts. He used the same argument to argue for God's existence since God was not evil nor deceptive and thus must exist. Nonetheless, he could have just as easily said, "Ego moventur, ergo sum" or "I move, therefore I am." This is because mind and body aren't different things, they are essentially one and the same. Derek Melser, the 21st century philosopher, argues from a neo-Rylean perspective that thinking is nothing more than the "covert tokening of concerted activity." That is, thinking is essentially embodied action. When we are thinking we are responding to the environment in some way such as reading, looking or doing and then reacting to it in some other way such as writing, talking or acting. We will return to Descartes' argument again when we examine the modern version of it in computer science: "A Brain in a Vat."

One might also question the role of the emotions in an embodied theory of thinking. The Harvard philosopher, Nelson Goodman, a cognitivist, once remarked that we probably rely more on feelings in our everyday lives than disembodied rationality.[6] To be sure, much of thought is bound up with our emotional life. For instance, as mental health clinicians have frequently pointed out, strong emotional arousal tends to narrow our range of attention. Attention is a cognitive process in which we (a) selectively focus on one thing in consciousness while we ignore other things, (b) divide our attention among several things or (c) sustain our attention over a period of time, that is, a hypervigilant attentional state. What we focus our minds on and think about is highly determined by our level of emotional arousal—as in deep interest in a topic or a sensation such as a pleasing aroma—as well as the range and kinds of emotional states we experience.

Similarly, Antonio Damasio, the neurologist, argues that bodily signals (visceral and nonvisceral sensations) modify the way the brain handles information, and this is accomplished primarily through use of our prefrontal cortices or lobes.[7] Because they lack these body signals or somatic markers, patients with damage to the frontal lobes are unable to anticipate the future. That is, they are unable to predict future events because they possess no biasing somatic state to rely on for their decision-making. Imagine yourself standing in a buffet line and having to decide between a piece of chocolate cake and a piece of apple pie. How does anyone typically make this type of decision? Certainly, there is no logical choice between the two if they are essentially the same price or calories. That is, there is no logical calculus that one can apply to arrive at deciding whether one option or another is preferable. Instead, we make such a decision based on our experience with these two food items from past experience and the pleasure we experienced consuming them. A creature without emotions would thus be helpless in the most mundane of everyday tasks. Captain Spock of Star Trek fame is a convenient dramatic fiction not a character that has any rational basis in science or even real life. Just like children that fail to survive because of a lack of ability to perceive and respond to pain, our fictional Spock, bereft of the ability to perceive and respond to emotions in himself and others, would be unable to make even the simplest of everyday decisions—often to avoid harm—or to form enduring relationships with caregivers and others essential for survival.

Instead, our preferences are typically bound up with our previous experiences. These previous experiences are associated in our minds with positive or negative emotional experiences and are inextricably bound to both our emotions and to our bodily states. If, as a frontal lobe patient, I cannot access these previous states, then there is no biasing somatic state to urge me to choose one food option, in this particular case, over another. I am in a decisional checkmate and like *Bartleby, the Scrivener*, in the novella by Herman Melville, unable to exercise any choices or options. While Bartleby made a conscious decision to refrain from choosing—"I prefer not to"—the frontal lobe patient is in a much more difficult predicament because she is unable to even place these choices in her decisional scratchpad in working memory, that is, place these options in consciousness and attend to them.

So, emotions are important components of an embodied mind but will serve as a backdrop when I am discussing topics that appear to uniquely arise from our emotional life and that science has shown clearly impact on the MBB including, (a) pathologies of the body that disrupt thought (autism, apraxia, and Parkinson's disease); (b) the contribution of the body to creative thought building on my own theory of early metaphoric development insofar as many types of creativity are deeply bound up with our emotional life; (c) thinking with the body through the arts and the use of the body in aesthetic expression and understanding such as music, dance, and the visual arts; and the (d) embodied nature of personality as well as the character of our communication with others.

The book is divided into ten major sections. In the first two sections I examine the neurobiological basis of the organization and mechanisms of thought, movement, and the body expanding on some of my own recent writing and theorizing on the subject. In section three, I investigate thinking with the body in non-human animals including an examination of the faculties of smell, sight, navigation, and object identification, which, I will argue, shed additional light on the biological and evolutionary basis of the embodied mind. Having a terrier companion at my side since he was 10 weeks old has surely been an asset. He has not only taught me deep humility toward *Canis familiaris*, but how much one can learn about and profit from the world from observing and interacting with other species. Conceptualization in science as an embodied mental activity where I investigate spatial cognition and mathematics makes up the fourth section. The disruption of thought in the pathologies of the body such as stuttering, Parkinson's and Huntington's diseases, as well as autism comprise the fifth sections of this book. Some of these diseases I have researched and studied, diagnosed or treated first hand, whereas others (Kallman's syndrome) I have had to rely on case histories and the scientific and medical literature.

In the fourth section, I elaborate on the contribution of the body to creative thought building on my own theory of the development of early metaphoric abilities in infants and young children. There is much confusion in the scientific literature about what even creativity and metaphor are—Is metaphor the kernel of creative thought?—and, briefly, I argue that creativity emerges out of a matrix of activities of the mind, brain, and body.

The arts have been central to my understanding of human thought and mentation and in section seven I look at thinking with the body through the arts, particularly in the areas of music and dance, and, to a lesser extent, the visual arts, areas which I know best. I also take up the use of the body in aesthetic expression and understanding. Is there a universal human aesthetic *sense* closely tied to the body or is appreciation of aesthetics an innate or learned ability for the privilege of the few?

In the eighth section, I exposit on the embodied nature of personality as well as the nature of linguistic and social communication with others. As it turns out, personality is not just in the head but distributed across the MBB. The role of movement, action, and activity in learning and pedagogy and the role of the embodied mind in sport and body therapies like the Pilates, Feldenkrais, and Alexander methods, as well as Yoga and other Eastern practices, is taken up in sections nine and ten. These skills and practices are important to the lives and hearts of tens of millions, so they deserve a reasoned examination.

In the final section I examine some of the theoretical issues that inform a mind embodied. For instance, language is not just about speech, but it involves communication with others through gesture, posture, facial expression, and vocal prosody, the overall system comprising body language or more properly understood as kinesics and proxemics. Whether speech emerged primarily out of bodily communication or social interaction or evolved separately but in parallel with gesture or with the intermediary of a protolanguage, are open questions. Some have argued that both speech and gesture emerged out of a more primordial matrix of the capacity for recursion, mental time travel, and theory of mind. In linguistics, recursion is classically defined as the ability to embed a noun phrase within another noun phrase or a clause within a clause. However, it looks just like the ability to embed one gesture or sub-gesture within another. It is this ability to think within a nested hierarchy of related concepts or ideas that is argued to be one unique feature of human thought. Therefore, it might be that this powerful mental module or a number of distinct interrelated modular elements evolved first in primates, and possibly other species such as dolphins, whose brain size is comparable to humans.

This putative collection of mental modules presumably provided the underlying cognitive architecture for the later emergence of speech and gesture, which emerged first with communication and thought centering on the hands and body, and later with communication and thought centering on speech in the evolution of modern hominins, that is, the human evolutionary lineage. Indeed, the brain systems that support both speech and gesture are similar and map overlapping areas in the human central nervous system (CNS). But that's not the whole story.

Let's look first at some design principles unique to the MBB in humans and other animals.

Notes

1. Stephen Hawking (1942–2018). Early in life after he had married and fathered children, he was severely incapacitated by a motor neuron disease that controls the voluntary muscles of the body. He ended up communicating through speech-generating software, a hand-held switch, and eventually just a single cheek muscle. A rare theoretical physicist and cosmologist.
2. Alberto Giacometti was a highly influential Swiss sculptor (1901–1966) who was known for his tall and slender figurines. His work was in the cubist-surrealist tradition that heralded the modernist movement and existentialism of the first half of the 20c, according to William Barrett in his book, *Irrational Man: A Study in Existential Philosophy* (1962). I read Barrett's book in college and visited the Museum of Modern Art (MOMA) in NYC, which has a large collection of his works. Giacometti was awarded the grand prize for sculpture at the Venice Biennale in 1962, a highly prestigious award. In 2015, his *L'Homme au Doigt* sculpture sold for $141.3 million at Christie's auction house in NYC. As a side note, the philosopher who introduced me to existentialism and pragmatism, including William James' writings, was Professor Bruce Wilshire of Rutgers (1932–2013), a noted authority on James.
3. The movie is *Invaders from Mars* (1953). A science fiction classic.
4. Deep Blue beat Gary Kasparov on May 11, 1997 in their second match (3 ½ to 2 ½). At the time, John Searle, a well-known philosopher at the University of California at Berkeley, when asked about the win said, "So what. Chess is an insignificant game" or something to that effect.
5. Gilbert Ryle (1900–1976) was a British philosopher who believed that *dualism* (the mind and body are separate elements) would never solve the riddle of the mind-body problem. Mental events were just the "ghost in the machine." Recent trends in the cognitive neurosciences, however, such as *embodied and situated cognition* have provoked a renewed attention to Ryle's ideas. He distinguished between *knowing how* or procedural knowledge and memory (how to do things) and *knowing that* or declarative knowledge and memory (knowing facts), which has had important influences on cognitive psychology.
6. One of the most important twentieth century philosophers of aesthetics, the Harvard philosopher, Nelson Goodman (1906–1998), wrote *Languages of Art: An Approach to a Theory of Symbols*. I read his book in graduate school after I had read Professor Howard Gardner's book on multiple intelligences (a mentee of Professor Goodman) and while I was doing research with Professor Harry Beilin, my dissertation advisor, on children's understanding of photography. My dissertation, however, was on the development of early metaphoric (creative) abilities in young children. Goodman's theory of notation became especially relevant when I later studied dance movement, choreography, and dance notation (labanotation) at the *Rudolf Laban Institute* and other venues in NYC.
7. Antonio Damasio, a physician-neuroscientist, has researched and written on music, emotion, and consciousness and how they emerge in the brain.

Bibliography

Damasio, A. R. (1996). The somatic marker hypothesis and the possible functions of the prefrontal cortex. *Philosophical Transactions of the Royal Society of London* (Series B), *351*(1346), 1413–1420.

Descartes, R. (1637). *Discourse on method and related writings*. Trans. D. M. Clarke. London: Penguin.

Goodman, N. (1968). *Languages of art: An approach to a theory of symbols*. Indianapolis, IN: Bobbs-Merrill.

Melser, D. (2004). *The act of thinking*. Cambridge, MA: MIT Press.

Melville, H. (2016). *Billy Budd, Bartleby, and other stories*. New York, NY: Penguin. (*Bartleby, the scrivener*, originally published 1853).

Ryle, G. (1949). *The concept of mind*. Chicago, IL: University of Chicago Press.

Seitz, J. A. (2018). Propaganda and war. *SOJ Psychology*, 5(2), 1–7. doi: http://dx.doi.org/10.15226/2374-6874/5/2/0015

· 1 ·

MIND, BRAIN, AND BODY

Design Principle 1—Dedicated System for Execution of Motor Acts

The brain is not an easy system to understand and our knowledge is limited in this area as so often the case is in areas of deep human interest. However, there are a number of design principles that we can articulate particularly as it relates to the motor system of the brain. For one, the execution of motor commands is primarily carried out in the motor cortex or motor strip (MC). Its location is easy to find (Figure 1.1). Simply take your finger and place it in the middle of the top of your head and draw it down to the top of your earlobe. The underlying neural tissue that comprises the motor cortex is roughly synonymous with this one-half inch wide area that lies in the rear of this frontal part of the cortex (hence, frontal cortex). If you decide to reach up and scratch your head, the execution of this movement is largely carried out by the motor cortex. So, design principle number one is that there is a dedicated system for the execution of motor acts that works in concert with other parts of the brain and body for the smooth execution of action.

As it turns out, lying to the back of (posterior to) the upper part of the motor cortex is an area named the premotor area (PMA) in which previously learned motor behaviors, such as the motor skills specifically used in pitching a ball, are on tap. It is here that previously learned actions are stored. If the skill is a new unlearned motor behavior, an area right below the premotor area, commonly referred to as the supplementary or lateral motor area (SMA), is responsible for acquiring or learning new actions. If you have never played golf or driven a stick shift on a car, the motor skills that you begin to acquire with practice and pluck are largely under the control of the SMA. Once learned, they are transferred to the PMA for quick usage and often performed without conscious thought. Indeed, most of what you actually do is carried out unconsciously by the brain. For instance, the cognitive mechanisms underlying the production of speech such as grammar, brushing your teeth, and executing a gesture to your pet dog to follow you. In the latter instance, you are conscious of your desire for your dog to accompany you, but you don't have to formerly think about what gesture to actually execute to get his attention. So, not just your emotional life is largely unconscious, as many of us have grown up believing, but much of what we think, say, and do. The former is attributed

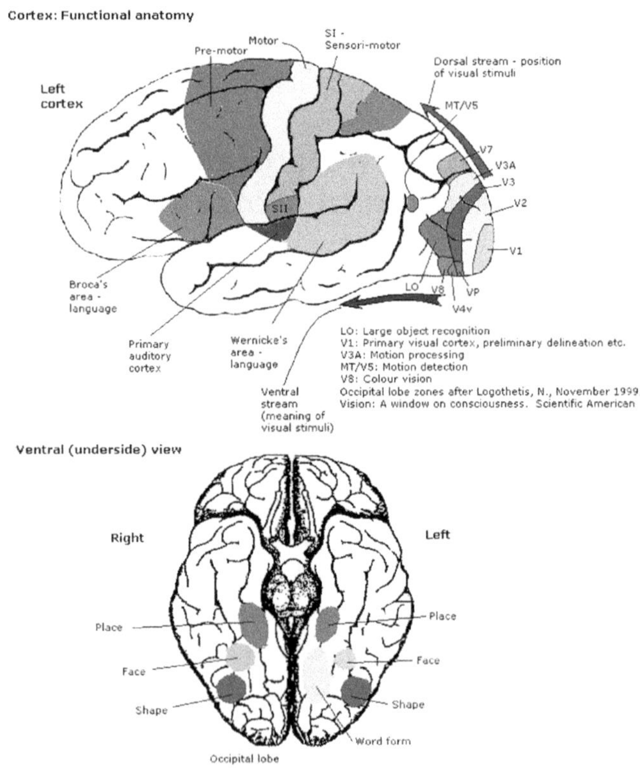

Figure 1.1. Human Cerebral Cortex.
Credit: Originally uploaded to Wikimedia Commons by RobinH at English Wikibooks. (https://commons.wikimedia.org/wiki/File:Constudproc.png), "Constudproc," https://creativecommons.org/licenses/by-sa/3.0/legalcode.

to the great Viennese neurologist, Sigmund Freud, who believed that much of our behavior is driven by our unconscious emotional life, that is, the emotional unconscious. The latter is part of what we think, say, and do and is part of the cognitive unconscious. In reality, the two significantly overlap.

In addition, each limb of the body contributes its own central pattern generators. These consist of ensembles of motor units that are made up of aggregates or pools of motor neurons as well as accompanying skeletal muscles under their control. For instance, imagine that you are performing a biceps curl with a small 10-pound dumbbell. Extending your arm with the dumbbell in hand is called a muscle extension. Flexing your muscle and pulling it back into your chest is known as a muscle flexion. The extension movement is performed by the triceps muscle that runs from your elbow to your shoulder on the outside of your arm. At the same time, the opponent muscles in your biceps that run from the indentation of your elbow to the top part of your shoulder are inhibited by these motor neuron pools. When you perform a flexion, the opposite obtains, the flexion movement is performed by the biceps as the opponent triceps muscles are inhibited. Pools of neurons work synergistically with skeletal muscles to carry this out.

Motor neuron pools and skeletal muscles for executing extension and flexion of muscle groups are further facilitated by motor maps in the motor cortices. The motor cortices, including the MC, PMA, and SMA, don't simply represent motor actions as sequences of behavior but are also in command of topographic maps of the layout of the body and their accompanying muscle groups. These motor maps must select which striated muscle groups—slow, intermediate or fast motor units—execute the action. All skeletal muscle tissue has a striated appearance because of a thick filament system of protein strands much like a bundle of electrical cables surrounded by a shielding or wrapper. Some of these collections of filaments have tremendous endurance but less force or power and are used to control muscle groups used in aerobic exercise. They are inundated with blood capillaries and used in such activities as walking, running, and bicycling and use muscle fibers known as slow-twitch muscle fibers. Other collections of filaments provide much more force or power but much less endurance and are used to control ensembles of muscles used in anaerobic exercise. They have a much smaller blood supply and are used in activities such as sprinting and weight-lifting and use muscle fibers known as fast-twitch fibers. And some fall in the intermediate range. As I mentioned above, the entire system for these and other head, torso, and limb movements are variously grouped together as *central pattern generators* or local control of large muscle groups.

The brain cannot do this alone because it needs information at the local level where the activity actually occurs as in speaking, gesturing, posturing or making facial expressions. By way of analogy, imagine you want to setup various web cameras to monitor your home and you want to be able to access this over the Internet at work when you are not at home. The computer or CPU (central processing unit) can't do this by itself but needs information at the local level such as access to web cameras to finesse this remote activity. These information devices will therefore need a "mind" of their own such as the ability to pan and rotate, orient to movement or light, engage in face detection as in some sophisticated systems, and so on. This indicates that our nervous systems must be distributed throughout our bodies to actually produce intelligent behavior.

Design Principle 2—Distributed System of Motor Control

An interesting property of the neuronal-muscular control of movement has been taken up by designers of robots and robotic vehicles. Rodney Brooks at MIT has designed robots from the ground up allowing for the robot to acquire or learn movement such as exploiting the ability to navigate over and around obstacles using a distributed system of learning and control rather than a more traditional system with a central organizer—brain or CPU—that issues commands and a body that follows blithely, slavishly or unthinkingly.[1] Something like the central pattern generators found in mammals described above. Therefore, design principle number two is that areas of the brain for the acquisition and storage of motor behaviors work in concert with the rest of the body to allow a distributed system of motor control. That is, motor control is not just in the head (brain).

Cephalopods, a class of invertebrates, particularly the octopus, have a distributed nervous system of about 500 million neurons, of which approximately 50 million are estimated to be in the cranium or *brain capsule*, about 150 million in the two optic lobes while the rest, approximately 300 million, are distributed among the four pairs of arms. These three systems, moreover, may function partly autonomously; the four pairs of arms can produce coordinated movements similar to other vertebrates without the intervention of the brain capsule or input from vision. In cephalopods, then, there is no obvious boundary between the mind-brain and the body. To be sure, in intact animals the three work intimately together.

Design Principle 3—Motor Abilities Are Directly Correlated With the Corresponding Amount of Brain Tissue

Another feature of the brain is that talents such as juggling, playing the violin, or driving a race car as well as specialized motor abilities such as knitting, dribbling a basketball, and executing an arabesque are directly correlated with the corresponding amount of cortical brain tissue that subserve each. Or, to put it another way, the more you use a body part or parts, as in fingers used in knitting, there is greater representation of that body part or parts in the cortex. And that is just another way of saying that the more you make use of a body part the brain allocates more brain tissue to the control of that body part and its activities. Indeed, there is mounting neuroscientific evidence for the close relationship between hand and mind. In right-handed string players who play the violin or cello there is a greater amount of brain tissue that is allocated for their left-hand digits—where the majority of musical action occurs in fingering the strings—than in non-musicians. Even when just listening to music, professional pianists exhibit involuntary motor activity such as tapping their fingers and that engages the primary motor cortex. More broadly, our perceptual experience—hearing music, feeling the wind over our bodies, smelling the aroma of fresh-brewed coffee or watching a gorgeous sunset—engages our bodily or sensorimotor knowledge and we naturally take advantage of this intrinsic relation between our hand and mind, our body and brain. Mental concepts may turn out, surprisingly then, to be nothing more than actional concepts. Concepts that inhere in our interactions with people, objects, events, and things in the world. As I type this, my fingers are part of the system that I think with. Emotions, too, are not just situated in the association areas of the brain and its specialized processing areas such as the amygdala and hippocampus, but as Candace Pert, one of the discovers of opiate peptides argues, are distributed throughout the brain and body.

There is a special name for the field that integrates some of these main components of the brain and body. It's called *psychoneuroimmunology*. It's the science of how psychological factors modify the functioning of the CNS in concert with the endocrine system—the system that houses the various hormones of the body—and the immune system that fights disease. The CNS itself is comprised of the brain and spinal cord. The brain cannot function alone, and many everyday actions are carried out, by structures in the spinal cord that enable basic reflexes. These are largely automatic actions under the control of groups of spinal neurons. When I move my hand away from a hot object or instantaneously jump out of the way of an approaching car, I am activating spinal cord-mediated responses. I do not have to think about these actions, that is, they are not facilitated by higher-order cognitive centers in the brain but are under the direct control of spinal cord neurons. Many actions, such as the motor behaviors I described above, are under local control and influenced by the body. This brings us to the role of the emotions in bodily actions.

Design Principle 4—Emotions Are Essentially Actions

My emotional state can influence what actions and what kinds of behaviors I engage in and the endocrine system and the immune system can modify the expression of these emotional states. The endocrine system consists of numerous structures including the hypothalamus, the pituitary gland, the thyroid gland, islets of Langerhans (part of the pancreas), the adrenal gland, the gonads (ovaries and testes), the pineal gland, and the thymus gland. These glands secrete endocrine hormones: Corticotropin-releasing hormone, adrenocorticotropic hormone, follicle-stimulating hormone, and growth hormone are secreted by the hypothalamus. Insulin and glucagons are secreted by the pancreas and glucocorticoids by the adrenal gland; testosterone, estradiol, progesterone, and thyroid stimulating hormone (TSH) by the pituitary gland; the thyroid hormones T3, T4, and calcitonin for calcium homeostasis by the thyroid; T lymphocytes by the thymus gland; and melatonin by the pineal gland.

Let's take one of these glands, the pancreas. The pancreas secretes a hormone (insulin) that controls the use of glucose by the body. The most common form of diabetes, Type 2, is the result of a lack of sufficient insulin production and the body's inability to make proper use of what is produced. But psychosocial stress increases the deposition of abdominal fat, a risk factor for diabetes, among other things. It appears, that this increased abdominal fat—at its extreme, obesity—predisposes people to Type 2 diabetes because of the actions of a group of hormones or cytokines (technically, adipokines) that are secreted by adipose (fat) tissue and modify the uptake of glucose into the cells of the body. Cytokines are proteins that make possible communication between the immune system and the CNS, and modulate the immune response, that is, how we respond to pathogens such as viruses, bacteria, and fungi as well as internal threats, such as tumors. Interestingly, family instability such as marital problems and job insecurity increases the incidence and severity of diabetes. In general, psychosocial stress such as lack of control on the job, family discord, and social isolation can lead to ailments such as depression, cognitive and memory impairments, cardiovascular disease such as hypertension (high blood pressure), as well as excess weight and poor sleep. One can therefore say that there is an intimate relation between the central nervous system, the body, and our external surroundings. Emotions, then, are part of the motivational system that energizes the organism to do something when confronted with challenges in the environment. Emotions are thus, essentially actions.

Design Principle 5—Shifting Actions Are a Form of Intelligence

With regard to the intricacies of movement, the basal ganglia and cerebellum have a tremendous influence on complex actions. The basal ganglia, a large clump of neurons, consists primarily of the caudate, putamen, and the globus pallidus in the "subcortex" or the part of the brain just below the cortex and deep inside the brain. It modulates movement initiation in feedback loops to the motor cortices. Individuals with Parkinson's disease have difficulty initiating movement because of the malfunction of two opposing pathways in the basal ganglia that facilitate one kind of movement (raising your arm) and, at the same time, inhibit contrary actions (lowering the same arm). These two opposing pathways include the direct or striatonigral pathway that facilitates the initiation of actions as well as their continuance over time. Walking is a good example as it's difficult for Parkinsonian patients to initiate movement and, once started, is difficult to stop. On the other hand, the indirect or striatopallidal pathway suppresses behaviors that interfere with walking, among many other motor activities, so that the current action is effectively carried out. The result in patients with Parkinson's disease is a quintet of symptoms: Akinesia or difficulty initiating movement, tremors in the limbs, bradykinesia or slowness of movements, cogwheel rigidity or jerkiness in limb movements, and postural instability. A basic feature of the nervous system is that "things get done" through the action of opponent processes. One structure facilitates, the other opposes. In the case above, both opponent processes originate in the basal ganglia. Huntington's disease, a progressive deterioration of the brain, has just the opposite effects from Parkinson's disease. Now, the basal ganglia instead of inhibiting movement, ratchets it up. Such patients uncontrollably flair their limbs and have awkward head and trunk movements. They are distinguished by a rare but inherited neurological disorder that results in *chorea* or dance-like movements that is characteristic of patients with Huntington's disease. But they have no conscious control of their uncomfortable attempts at the oldest of the terpsichorean arts.

These dual circuits suggest that the ability to initiate new actions—that is, shift from an old, overused behavior to a new, underused behavior—underlies the creation of novel actions or the combining of old behaviors into new and unique sequences. This fundamental aspect of the brain and body suggests that *shifting actions*, the ability to create new sequences of behavior, underlies intelligence. And not just in humans but in non-humans, such as your everyday assortment of diverse canines and felines as well as other animals, too.

Of even greater importance than the basal ganglia, is the cerebellum, a massive structure sitting off the back of the cortex and comprising approximately fifty percent of all brain neurons. It guarantees smooth and efficient movement, moment-to-moment, by reducing errors between intended and actual movements used in complex motor sequences. Recently, however, reciprocal and parallel neural pathways between the cerebellum and the frontal cortex have been discovered. The most evolutionarily advanced part of the cortex, the frontal area is involved in conscious working memory or what you are thinking and pondering right now. These are so-called *executive functions* or the mental monitoring, planning, and managing of things you're currently thinking about as well as focusing your attention as you work through these diverse mental tasks. Such mental tasks might include reading this text, writing a letter to your aunt or speaking to a friend. Why might these reciprocal and parallel pathways exist?

Design Principle 6—We Think Kinesically

Contemporary research findings in the neurosciences indicate reciprocal and parallel pathways between the cerebellum and the frontal cortex suggest the close association between thought and movement. Are thought and movement so inextricably intertwined that we literally think kinesically? That is, do we literally think with our bodies?

For instance, the ability to shift from one mental task to the next is one function of the basal ganglia and, as I described above, is also the part of the motor system that malfunctions in Parkinson's and Huntington's diseases. This suggests that the same system that initiates movement is also responsible for shifting among mental tasks. Imagine that you are feeding an infant at the dinner table and she drops the milk bottle onto the floor. At the same time, she starts to mildly regurgitate. You have to immediately stop what you are doing and attend to the baby's needs as well as kneel down and pick up the oozing bottle on the floor. You have to change up on your own immediate routine and switch to a new set of mental tasks and actions with novel components. Of the latter, how do I get my infant to stop throwing up? Where did that bottle of milk land? What is interesting here is that a malfunctioning basal ganglia leads to stereotyped movement patterns and the absence of novelty or novel behaviors as occurs in such disorders as obsessive-compulsive disorder (OCD) and Gilles de Tourette syndrome. We will take up both of these disorders below.

The basal ganglia are part of two major cortical-subcortical circuits in the brain, that is, the cortical or thinking part of the brain works together with the older subcortical or non-thinking part of the brain to accomplish many diverse mental tasks. This distinction, however, is somewhat arbitrary because the two systems, old and new, are both needed to function together for human and non-human mentation to properly operate.

Here are the two cortical-subcortical circuits. The lateral orbitofrontal circuit (LOFC) is a circuit in the front of the brain, above and to the edges of the center of the cortex, that facilitates the ability to shift from one mental set to another. The dorsolateral prefrontal circuit (DPFC) is a circuit in the front of the brain, below and to the sides of the center of the cortex, that stores mental representations or what cognitive scientists define as internal models of the world. These internal models include what we imagine about the beliefs and feelings of others, how things and objects are constructed and function, how events unfold, as well as our latent knowledge of spatiotemporal contexts concerned with orientation and position of our bodies in three-dimensional space.

Moreover, the partial collateral storage of these internal models of the world in the basal ganglia may help the frontal cortex in implementing appropriate behaviors given the context of one's current experience. To take an earlier example, let's say that again we're both standing in a food line and chatting and while speaking have to make some decisions about what we are going to eat. The LOPC allows us to switch back-and-forth from conversation to grabbing a food item while the DPFC allows us reach and grab that particular piece of chocolate cake in space and time even in mid-sentence. The breakdown of these circuits has interesting repercussions on behavior. Let's look at the breakdown of these circuits by first examining some important functions of the historically misunderstood cerebellum.

As it turns out, the cerebellum has some very important non-motor or *cognitive* functions. Historically, it was thought to be just a large motor structure that coordinated all of the major muscle groups in the body including the musculature for speech. Words can be spewed out by modern humans at a rate of up to 140 words-a-minute, which requires complex coordination of both the speech articulatory apparatus as well as careful attention to explicit timing of the serial packaging of phonemes to create word forms. The structure and function of the cerebellum, however, leads one to believe that thinking and acting are two sides of the same coin. For one thing, the cerebellum is known to play an important role in timing functions. For instance, typing on a keyboard, walking down the street, playing the piano, and picking up a cup to sip; yes, picking up a cup in time and space involves coordination and timing. But there is also recent evidence that it, along with the basal ganglia, play a role in planning, regulation, and attention. There are two separate pathways for planning movement: the extrapyramidal tracts originating in the brainstem and the corticospinal tracts originating in the motor cortex. The extrapyramidal tracts specify the goal or target location of an action as in throwing a baseball to a catcher behind the mound, whereas the corticospinal tracts specify the trajectory of action as in throwing a curved ball.

It has been shown, moreover, that the cerebellum contributes to basic associative learning processes such as associating words with colors and the ability to rapidly shift attention within and between sensory modalities. When visually identifying an object such as a banana, for example, you may find yourself perusing its color and shape for ripeness and then rapidly switching your orientation to its piquant odor to confirm your initial belief. Nonetheless, autism, an inability to relate emotionally and socially to others, has been hypothesized to occur when maldevelopment of the cerebellum leads to poor social and cognitive development. This is true even in the absence of damage to the hippocampus and amygdala, which are brain structures known for their emotional and motivational importance including the processing of affective and social memories. We will look at Autistic Spectrum Disorders (ASD) in Chapter 3.

Motor Activity as the Basis of Human Intelligence

There is evolutionary evidence that more advanced areas of the cerebellum including the neocerebellum and the dentate nucleus as well as kindred and cognate regions of the frontal lobe evolved in parallel in hominids. Ancestors of modern humans (hominins) included the Great Apes (hominids): Archaic humans, orangutans, bonobos, gorillas, and chimpanzees. The evolution of the neocerebellum enabled the integration of features of modern human mentality as in the use of stereoscopic vision, that is, seeing in three-dimensions, and the complementary and accompanying use of the hands to manipulate three-dimensional objects.

One might thus argue that motor activity is the basis of human intelligence. Indeed, it has been postulated that the core of human intellect is the capacity of more recent abilities such as our cognitive or computational domains to draw on abilities that evolved for other tasks such as motor skills or abilities. Cognitive and motor behaviors therefore got hooked up over evolutionary time. How might have this occurred?

Origins of Mind

The origins of mind are a speculative area. But biologists have offered some plausible evolutionary scenarios. The following is one suggested by the neurobiologist, M. Deric Bownds. The first thing you need is a self-replicating molecule and DNA satisfies this admirably as a universal Darwin machine. That is, DNA, because it is self-replicating, is able to generate new biological forms. Canines are a good example because they evolved from wolves and have many variants such as border collies for herding and pointers for recovering game. These new biological forms and their variants are *tested* for and selected within some environment and those forms and variants most *adaptable* within their present environment (not *fittest*) are more likely to pass on their DNA by way of their genes to their offspring and future generations.

Initially, the simplest forms of behavior evolved and included the ability to sense chemicals in the environment as well as incorporated cell control machinery in the form of a cell nucleus. These were the beginning of the *eukaryotes* although earlier one-celled organisms, so-called prokaryotes, lacked a nucleus. These chemical sensors facilitated the beginnings of early approach and avoidance behavior as well as obtaining needed nutrients in the environment and avoiding toxic ones. One might say that the evolution of cell nuclei enabled individual cells to begin to start *processing information from the external and internal worlds* for themselves. With more control at the local level, cells could begin to communicate with each other as well as link up with other cells to form multicellular circuits. These nascent multicellular circuits begat the simplest of behaviors and actions, commonly referred to as *reflexes*. Reflexes are classically defined as *involuntary behavioral responses* and are automatic responses of cells to changes in the external environment. For instance, moving away from the scorching heat of the sun. Once you have a reflex circuit as well as the ability to sense chemicals in the environment you have effectively separated sensation from action. Sensation is the conversion of physical stimulation impinging on the organism from the outside into electrochemical information that is processed inside the cell by the cell machinery: Nuclei, mitochondria, and chloroplasts for converting energy garnered from the outside milieu, and various organelles for interior *housekeeping*. Action is nothing other than another word for behavior and is accompanied by various movements and activities facilitated by these multicellular circuits.

Now, with the establishment in evolution of separate sensory receptor (sensory) and effector (action) circuits—that is, one circuit for sensing the external environment and one circuit for responding to these sensations—you would have needed a way to hook up these components and an intervening cell was one solution, called an *interneuron*. The latter arose to coordinate information from the sensory circuits with the resultant output or actions that emanated from the receptor. Eventually, this interneuron proliferated and connected up with other interneurons to enable a representation of an action. Only if you can store and recall past experiences with the environment can history offer the organism any barometer of what to do under similar circumstances in the future. And such an ability would be the first occurrence of a *thinking body*, a *mind-brain-body* (MBB), or a *mind embodied*. Down the evolutionary road, these so-called *representations* increased in number and diversity and their use could then be more effectively tailored to the appropriate occasion.

Likewise, the sensory side of these circuits may have given rise to two kinds of preferences by the organism: Increasing physical stimulation from the outside or decreasing it. These preferences may have arisen by way of evolution of a feedback loop embedded in these sensory circuits themselves that was self-propagating. Thus, preferences for desirable things were self-perpetuating while preferences for undesirable things were self-negating. The former is positive and enhances feedback from the environment while the latter is negative and inhibits feedback. We call this evolutionary circuit *affect*.

Another Set of Dual Circuits: Sensing the Internal and External Worlds

Later in evolutionary history a dual circuit arose that enabled the organism to distinguish between sensation arising from within the organism and sensation transduced from the outside world. This dual-mode of two distinct mental models that evolved to distinguish between internal sensation and external perception turned out to be a very useful development for evolving organisms. Imagine if you could not differentiate between your own thoughts and the digital prognostications coming from your television or computer. Indeed, individuals with a rare hereditary disorder believe, in some cases, that the voice of the broadcaster or the spoken words of others are actually emanating from inside their own heads. This symptom, one of many, has been referred to historically as *thought broadcasting*. The disease for which it is typically associated, schizophrenia, is a devastating neurological and mental disorder that co-ops the brain's ability to effectively process internal communication resulting in various kinds of *thought disorders* as well as distinguish between internal and external communication with the outside world. It is, in effect, a breakdown in the brain's information-processing capacities. It's sort-of like your usually trusty laptop computer that accesses the wrong file or a word-processing application that has trouble communicating with the kernel of the computer chip and the notebook *locks up* and you have to re-boot the computer. In the case of individuals with psychosis, such as schizophrenia or other diseases of the mind, new chemical as well as electrochemical therapies as they are called, psychotropic drugs, and other psychopharmacological agents effectively perform the same function. They get the brain circuits communicating again. I'll have more to say on this later.

An example of these dual circuits is the evolution of the vertebrate eye. In this case, organisms developed separate channels for both visual sensation and visual perception. With this ability, one could visually scan the environment for information and then compare it to previously stored visual information in memory. A great device, as it turns out, to obtain much-needed nutrients and avoid predators. Obtaining food and other desirable resources like mates and places to nest depended on this dual circuit (and similar ones tuned to sound, touch, taste, and smell) and made predictions about the future feasible: Where are those tasty berries I stumbled over last week? Once the ability to compare the past with the present and anticipate what might happen next became possible, an organism could predict the behavior of other organisms. For instance, why is that animal looking at me?

Posterior Parietal Cortex

The influence of the brain on movement is not just found in the motor cortices (MC, PMA, SMA), basal ganglia, and cerebellum but in other areas of the cerebral cortex. Another so-called motor structure, the posterior parietal cortex (PPC), helps create a frame of reference for movement. It comprises the upper sides and top of the cortex and lies towards the back of the brain. The frame of reference created by the PPC includes movement in three-dimensional space abetted by visual, vestibular or haptic feedback. That is, we might receive feedback from our eyes about the movement of our hand as reach for a plate in the cupboard (visual and spatial feedback), but the feedback might just as well be from our sense of balance and equilibrium as we reach for an article of clothing high on a shelf standing on a chair (feedback from the vestibular organs of the inner ear) or reach way in the back of a closet for a wool garment we have not worn since last winter (largely relying on touch or haptic feedback). The posterior parietal cortex takes this sundry feedback from the aforementioned senses and routes it to the appropriate areas of the brain concerned with sensory motor imagery. Areas of the temporal cortex on the left and right sides of the outer layer or *mantle* of the brain for visual mental imagery, putatively the insular cortex for vestibular imagery, and the somatosensory cortex for tactile imagery. The insular cortex is folded deep within the Sylvian fissure that separates the temporal, parietal, and frontal lobes. The somatosensory cortex (technically, Brodmann's areas or BA1, 2, & 3) is located in the postcentral gyrus across the top of the middle of the head if you have a finger handy. Let's now look at the big picture.

Motor Logic, Motor Organization, Kinesthetic Awareness, and Kinesthetic Memory

It should be clear from the following discussion so far that motor functions are distributed throughout the brain and body. We can group these motor functions into broader categories of interacting systems of brain and body that I will subsequently refer to as motor logic, motor organization, kinesthetic awareness, and kinesthetic memory. For instance, new evidence on the function of the motor cortex (MC) indicates that it stores and implements the serial order of a motor plan. For example, cooking dinner comprises several or more steps that are carried out in a particular order. Memory for those steps, or what is called *kinesthetic* or *procedural memory*, is stored in the motor cortex as well as the executable "files" that actually carry out the plan. That is to say, there is both memory for the individual steps as well as their overall sequence. This is referred to as the *syntax of movement* or what I and others have called *motor logic*. The organization of this movement so that the parts work together in space and time is referred to as *motor organization*. So, movement, action, and activity have an organization and a logic. The organization is based on motor patterns held in memory that are returned as a sequence of motor actions. Cooking dinner is a good example because it evolves a series of actions such as preparing raw ingredients, buttering the frying pan, and turning on the stove that are nevertheless constrained by the context of the actions. You must prepare the ingredients and often in a particular order before actually cooking them. Motor logic, however, originates in the musculoskeletal system because the body can only carry out certain actions in specific ways. There may be many variations on a golf swing, but the basic constraints of the musculoskeletal architecture and biomechanics of the human body determine a fairly precise range of limits on any specific swing. This is the *logic of the body* that is often spoken about by dancers; athletes; billiard players; movement specialists, that is, practitioners of Alexander, Feldenkrais, Pilates, and other assorted movement therapies; massage, physical, and occupational therapists; physiatrists and orthopedic surgeons; and the like.

In analogy to formal logical propositions, any action has a set of formal biomechanical constraints determined by the musculoskeletal architecture of the body such that any movement is limited by both its underlying bodily structure and the set of actions that emerge from this structure over time. An example is the movement of the arms within the shoulder rotator cuff as only certain actions are permissible or possible and not others. These are the formal propositions or facts about the body. A series of actions would link a set of biomechanical programs that emerge from the body and would only be valid if any set of connected actions makes appropriate transitions from one movement to the next. These are the valid inferences that emerge from the linkages of any constrained set of movements or actions; the results of the set of movements being what is known as a conclusion of a *logical sequence*. So, we could say that the body has its own natural logic.

Sequencing

Because movement always unfolds as a sequence in time and space, it has been argued that there is a general-purpose brain module for sequencing that participates in both movement and other cognitive tasks such as language, thinking, and so on. Sequencing, however, is not just a feature of making a series of neural linkages in the brain over time. Serial orders that emerge in behavior may actually occur in parallel because the brain is doing many things at once. Indeed, there is a distinction in computer science between *multiprocessing*—basically multitasking or doing many things at once—and *multithreading*. A mind/body or a computer that can do many things at once, that is, run multiple programs in parallel can also dedicate many underlying brain systems or computer programs to a complicated task at the same time or in parallel. This is known as multithreading in "computerese." In actual human action, playing a game of tennis comes to mind.

So, we have dedicated systems of the brain that are specialized for different intellective functions and these specialized systems can act in unison to complete a complex and intertwined set of mental tasks as in managing the complex operations of a business enterprise. Nonetheless, actions emerge over the breadth of moments and this logic of the body and the organization of movement over time is restricted by both the body's musculoskeletal architecture as well as forces that operate externally on the body such as other objects, gravity, wind, and so on. As a result, sequencing is not just in the head. Can a child have a sequencing disorder? Not in the sense discussed here. A problem in sequencing words would be a very serious but extremely rare disorder and there is some recent evidence that it is controlled by a gene (FOXP2), or much more likely a set of genes, and that it arises in certain highly isolated family groups. The disorder, known as *developmental verbal dyspraxia* or childhood apraxia of speech, is a result of an inability the MBB to work together with the appropriate parts of the body—the motor articulatory apparatus—to plan and sequence word sounds. The core impairment, then, is programming spatiotemporal parameters of sequenced movement and linking those sequences to appropriate sound production and vocal prosody of the body.

On the other hand, difficulty in sequencing movements or motor behaviors in young children would most likely represent immature maturation of motor control of the body that would, in most cases (the exceptions would be pathological such as the syndromes of cerebral palsy), be temporary and wane with age. The developing child has a normally complicated task ahead of him or her. Not only is there maturation of the underlying neural architecture that unfolds over development and helps organize movement, but the child's body must acquire the ability to control complex movements within a penumbra of constraints. These include the pull of gravity on the body; accommodating to the shapes, weight, and sizes of objects including people, as well as other internal and external forces that operate on the body. A short list of these would include the constraints of clothing and shoes; external environmental forces such as wind, rain, slippery pavement; and internal physical forces such as lateral acceleration. Once a body or body part is in motion its movement must be shaped, delimited or inhibited and acquiring these *motor skills* takes many, many years of learning and practice. Biology is not enough. It reminds me of an oft-quoted joke. A young man carrying a violin case is walking down a busy Manhattan street and is stopped and asked by another fellow, "How do I get to Carnegie Hall?" The young man responds, almost without pause, "Practice, practice, practice." Thus, the underlying neural hardware is shaped by the actions of the body and experience; it is not set at birth. The resulting collective actions are stored in the MBB as kinesthetic or procedural memory, that is, memory for how to do things. When it is called up for immediate use in some ongoing situation—for instance, prying open a stuck lid of jam or jelly—it is called *kinesthetic awareness*. Therefore, motor logic, motor organization, kinesthetic memory, and on-line kinesthetic awareness form the core of the intelligent operation of the motor system. Sensory systems, nonetheless, guide thought and movement. The nervous system does not so much direct behavior as shape the dynamics of the coupled system of brain, body, and environment. Movement doesn't exist on its own but is actually situated among musculoskeletal, environmental, cultural, and social forces. That is, brain and environment form a synergistic relationship in which brain events are embedded in a social and cultural matrix. In the field of mental health, for example, many symptoms of anxiety and depression originate in social distress and when the social and cultural stressors are lifted, the symptoms disappear. Mental processes, even dysfunctional ones, are shaped by the surrounding social and cultural environment. They are prepared by evolution—that is, evolutionary processes shaped by experience and the environment—and have often evolved to solve certain problems in living. For instance, depression may have evolved as a way of obtaining help from others.

Movement Initiation

There is another important aspect of movement; its initiation. The ability to decide to move and move in a certain way is part of a complex system of voluntary movement selection and initiation. Parkinson's disease is fundamentally marked by the inability to initiate movement. Parkinson's has classic and telling symptoms to jog your memory from our earlier discussion: Impaired imbalance, slowness of movement, "cogwheel" rigidity or stiffness in moving one's limbs due to heightened muscle tone, and tremor in the limbs principally before movement is initiated. Where in the brain does one find the official headquarters for movement initiation? As it turns out, one part is deep in the brain and is called the cingulate cortex. Its rostral or frontal portion is innervated by—has connections to—the amygdala and to the ventral or underside portion of the striatum.

The amygdala is important because it is the involved in the initial laying down or encoding of emotional memories whereas the ventral striatum appears to play a role in addiction or pleasure in reward. Addiction, of course, is not just a brain process but involves a set of behaviors and actions that are continually repeated in relation to some self-reinforcing act that is experienced as pleasurable when it is rewarded. Gambling, drugs, and sex are prototypical instances.

The ventral striatum and its accompanying structures have long been seen as the boundary between the emotional and motor systems of the brain. Whereas the anterior or rostral cingulate area connects to the emotional processing areas of the brain, the caudal or posterior portion of the cingulate area innervates the major motor areas of the cortex or thinking part of the brain: The primary and secondary motor areas or MC, PMC, SMA cortices. It also connects to the brainstem, which is one of the major conduits to the cerebellum or the brain's digital computer for motor functions. And it receives inputs from the frontal cortex, which handles the difficult task of managing all those different and continuous events that are popping off in the rest of the brain. Keeping everything in order and flowing properly, is one of its major tasks.

Primary, Secondary, and Tertiary Areas of the Brain

Let's also make clear the distinction between primary and secondary areas in the brain. Primary areas, in general, are where initial processing of specific sensory attributes occurs whereas *secondary* and *tertiary* areas, which spatially surround the primary area, fill out the cognitive, perceptual or sensory picture, so to speak. Imagine focusing your eyes on an object that you can't quite make out on a table, 30 feet away. Initially its color or hue and the lines it carves in space stand out (primary sensory area in the occipital lobe or visual cortex), and as you discern its shape and stem line at the top and small leaf-shape object (secondary sensory area), you recognize it as a three-dimensional and familiar object or an apple (tertiary sensory area). This overlaps with the general distinction made among the terms, sensation, perception, and cognition. Sensation occurs at the level of the senses such as the receptors in the eye that convert electromagnetic energy (visible light) into electrochemical neural impulses. At the level of the brain, the primary areas start to process this neural information (perception) and the secondary and tertiary areas flesh it out into a recognizable object that you can name and describe in other parts of the brain (cognition).

Body Schema

One other important function that the brain carries out is a mental layout of your body. What's commonly referred to as your "body schema" is really your mental representation or mental self-knowledge of the spatial picture of your body. Very useful information as you scratch your back for that intolerable itch that you can't see.

As it turns out, there is a distributed neural network or collection of brain regions that form the neural basis of corporeal awareness. It includes the touch or somatosensory cortex, the limbic or emotional cortex deep within the brain, and other sensory cortices that process information from the different sensory receptors about features of your body: Its scent, spatial location of parts, dermal sensitivities, texture, taste (your thumb, maybe, when you were a child), its auditory properties (try tapping on the top of your head), and so on. The visual cortex, for instance, processes what you see and what you see forms one piece of evidence that your brain uses to *understand* your body; you begin to form a cognitive map or spatialized layout of your body with your eyes beginning from the moment of birth. Human infants use this information to begin to map equivalences between the bodies of others and their own bodies.

Even so, because sensory systems mature at different rates, early reaching in infants is actually first channeled through kinesthetic or proprioceptive input. This internal feedback from specialized receptors in our muscles and tendons guides movement and is known as proprioception. Only later in the first year of life does visual and auditory guidance become important as well as *expproprioceptive* information from the movement of body parts in relation to the environment and *exteroceptive* information from the actual physical layout of objects and surfaces in the infant's immediate environment. Infants quickly discover that sound gives clues to object movement and location and that you can follow a moving object more easily with your eyes, which is an example of expproprioceptive information. And they learn that they must accommodate their movements to people and things—or collide—and that surfaces, like sand, constrain movement, which is an instance of exteroceptive information.

Thus, we have these amazing interactions of mind, brain, body, and environment. Some of these amazing properties of the body we will look at next.

Note

1. Rodney Brooks, formerly a Professor of Robotics at MIT, is now the Chairman and Chief Technology Officer at Rethink Robotics, which is working on collaborative robots for manufacturers.

Bibliography

Andreasen, N. C., Paradiso, S., & O'Leary, D. S. (1998). "Cognitive dysmetria" as an integrative theory of schizophrenia: A dysfunction in cortical-subcortical-cerebellar circuitry? *Schizophrenia Bulletin*, *24*(2), 203–218.

Berlucchi, G., & Aglioti, S. (1997). The body in the brain: Neural bases of corporeal awareness. *Trends in Neurosciences*, *20*(12), 560–564.

Bertenthal, B. I. (1996). Origins and early development of perception, action, and representation. In J. T. Spence, J. M. Darley, & D. J. Foss (Eds.), *Annual Review of Psychology* (vol. 47, pp. 431–459). Palo Alto, CA: Annual Reviews.

Bertenthal, B. I., & Clifton, R. K. (1998). Perception and action. In W. Damon, D. Kuhn, & R. Siegler (Eds.), *Handbook of child psychology, Vol. 2: Cognition, perception, and language* (5th ed., pp. 51–102). New York, NY: Wiley.

Bownds, M. D. (1999). *The biology of mind: Origins and structures of mind, brain, and consciousness.* Bethesda, MD: Fitzgerald Science Press.

Bracke-Tolkmitt, R., Linden, A., Canavan, A. G. M., Rockstroh, B., Scholz, E., Wessel, K., & Diener, H.-C. (1989). The cerebellum contributes to mental skills. *Behavioral Neuroscience*, *103*(2), 442–446.

Brooks, R. A. (1999). *Cambrian intelligence: The early history of the new AI*. Cambridge, MA: MIT Press.

Bushnell, E. W., & Boudreau, J. P. (1991). The development of haptic perception during infancy. In M. A. Heller & W. Schiff (Eds.), *Psychology of touch* (pp. 139–161). Hillsdale, NJ: Lawrence Erlbaum.

Carpenter, A. F., Georgopoulos, A. P., & Pellizzer, G. (1999). Motor cortical encoding of serial order in a context-recall task. *Science*, *283*, 1752–1757.

Chiel, H. J., & Beer, R. D. (1997). The brain has a body: Adaptive behavior emerges from interactions of nervous system, body and environment. *Trends in Neurosciences*, *20*(12), 553–557.

Cote, L., & Crutcher, M. D. (1991). The basal ganglia. In E. R. Kandel, J. H. Schwartz, & T. M. Jessell (Eds.), *Principles of Neural Science* (3rd ed., pp. 647–659). Norwalk, CT: Appleton & Lange.

Curran, T. (1995). On the neural mechanisms of sequence learning. *Psyche*, *2*(12), [online]. http://psyche.cs.monash.edu.au/v2/psyche-2-12-curran.html.

Daruna, J. H. (2012). *Introduction to psychoneuroimmunology*. (2nd ed.). San Diego, CA: Elsevier.

DeLong, M. R. (1990). Primate models of movement disorders of basal ganglia origin. *Trends in the Neurosciences*, *13*(7), 281–285.

Elbert, T., Pantev, C., Wienbruch, C., Rockstroh, B., & Taub, E. (1995). Increased cortical representation of the fingers of the left hand in string players. *Science*, *270*(5234), 305–307.

Fiez, J. A., Petersen, S. E., Cheney, M. K., & Raichle, M. E. (1992). Impaired non-motor learning and error detection associated with cerebellar damage: A single case study. *Brain*, *115*, 155–178.

Freud, S. (1940). *An outline of psychoanalysis*. London: Penguin.

Gallagher, S. (1986). Body image and body schema: A conceptual clarification. *Journal of Mind and Behavior*, *7*, 541–554.

Gardner, E. P., Martin, J. H., & Jessell, T. M. (2000). The bodily senses. In E. R. Kandel, J. H. Schwartz, & T. M. Jessell (Eds.), *Principles of neural science* (4th ed., pp. 430–450). New York, NY: McGraw-Hill.

Gazzaniga, M. S., Ivry, R. B., & Mangun, G. R. (2014). *Cognitive neuroscience: The biology of mind*. (4th ed.). New York, NY: W. W. Norton.

Ghez, C. (1991a). Voluntary movement. In E. R. Kandel, J. H. Schwartz, & T. M. Jessell (Eds.), *Principles of Neural Science* (3rd ed., pp. 609–625). Norwalk, CT: Appleton & Lange.

Ghez, C. (1991b). The cerebellum. In E. R. Kandel, J. H. Schwartz, & T. M. Jessell (Eds.), *Principles of Neural Science* (3rd ed., pp. 626–646). Norwalk, CT: Appleton & Lange.

Godfrey-Smith, P. (2016). *Other minds: The octopus, the sea, and the deep origins of consciousness*. New York, NY: Farrar, Straus & Giroux.

Goldberg, G. (1985). Supplementary motor area structure and function: Review and hypotheses. *Behavioral and Brain Sciences*, 8, 567–616.

Greenfield, P. M. (1991). Language, tools and brain: The ontogeny and phylogeny of hierarchically organized sequential behavior. *Behavioral and Brain Sciences*, *14*(4), 531–595.

Haueisen, J., & Knosche, T. R. (2001). Involuntary motor activity in pianists evoked by music perception. *Journal of Cognitive Neuroscience*, *13*(6), 786–792.

Heller, M. A., & Ballesteros, S. (2006). *Touch and blindness: Psychology and neuroscience*. Mahwah, NJ: Lawrence Erlbaum.

Kolb, B., & Whishaw, I. Q. (2015). *Fundamentals of human neuropsychology*. (7th ed.). New York, NY: Worth.

Lashley, K. S. (1951). The problem of serial order in behavior. In L. A. Jeffress (Ed.), *Cerebral mechanisms in behavior: The Hixon symposium* (pp. 112–136). New York, NY: Wiley.

Lockman, J. J., & Thelen, E. (1993). Developmental biodynamics: Brain, body, behavior connections. *Child Development*, *64*(4), 953–959.

Lopez, O. (1999, June). *The relationship between nonverbal behavior and thought.* Senior Honors Thesis, NIH/Minority Access to Research Careers (MARC) Program. New York, NY: York College/City University of New York, Department of Political Science and Psychology.

Lopez, O., & Seitz, J. A. (1998, November). *The bodily basis of thought.* Paper presented at the Annual Meeting of the National Minority Research Symposium, New York, NY.

Magill, R. A. (1989). *Motor learning: Concepts and applications.* (3rd ed.). Dubuque, IA: William C. Brown.

Mirani, C., & Seitz, J. A. (1998, August). *Development of body-gestural skills in children.* Paper presented at the New York Academy of Sciences, New York, NY.

Morgan, J. E., & Ricker, J. H. (2008). (Eds.). *Textbook of clinical neuropsychology.* New York, NY: Taylor & Francis.

Pennington, B. F. (2009). *Diagnosing learning disorders: A neuropsychological framework.* (2nd ed.). New York, NY: The Guilford Press.

Pert, C. D. (1999). *Molecules of emotion: The science behind mind-body medicine.* New York, NY: Scribner.

Purves, D., Augustine, G. J., Fitzpatrick, D., Hall, W. C., LaMantia, A. S., McNamara, J. O., & Williams, S. M. (2004). *Neuroscience.* (3rd ed.). Sunderland, MA: Sinauer Associates.

Ryle, G. (1949). *The concept of mind.* Chicago, IL: University of Chicago Press.

Seitz, J. A. (1993, March/April). I move … therefore I am. *Psychology Today*, *26*(2), 50–55.

Seitz, J. A. (1994, June). *Thinking kinesically: Theory and practice.* Paper presented at the 24th Annual Symposium of the Jean Piaget Society, Chicago, IL.

Seitz, J. A. (2000a). The bodily basis of thought. *New Ideas in Psychology: An International Journal of Innovative Theory in Psychology*, *18*(1), 23–40.

Seitz, J. A. (2000b). *Embodied cognition.* Paper presented at the Annual Meeting of the American Psychological Society, Miami, FL.

Stein, B. E., & Meredith, M. A. (1993). *The merging of the senses.* Cambridge, MA: MIT Press.

Sternberg, R. J., & Sternberg, K. (2012). *Cognitive psychology.* (6th ed.). Belmont, CA: Wadsworth.

Thelen, E. (1995). Motor development: A new synthesis. *American Psychologist*, *50*(2), 79–95.

Turvey, M. T. (1990). Coordination. *American Psychologist*, *45*(8), 938–953.

White, N. M. (1997). Mnemonic functions of the basal ganglia. *Current Opinion in Neurobiology*, *7*(2), 164–169.

Wickelgren, I. (1998). The cerebellum: The brain's engine of agility. *Science*, *281*(5383), 1588–1590.

Wickelgren, I. (1999). Memory for order found in the motor cortex. *Science*, *283*(5408), 1617–1619.

Yeates, K. O., Ris, M. D., Taylor, H. G., & Pennington, B. F. (2010). (Eds.). *Pediatric neuropsychology: Research, theory, and practice.* New York, NY: The Guilford Press.

Youdim, M. B. H., & Riederer, P. (1997). Understanding Parkinson's Disease. *Scientific American*, *276*(1), 52–59.

· 2 ·

THE BODY AS AN ACCESSORY TO THE MIND

Now it is the opinion of Anaxagoras that the possession of these hands is the cause of man being of all animals the most intelligent. But it is more rational to suppose that his endowment with hands is the consequence rather than the cause of his superior intelligence. For the hands are instruments or organs, and the invariable plan of nature in distributing the organs is to give each to such animal as can make use of it; nature acting in this matter as any prudent man would do. For it is a better plan to take a person who is already a flute-player and give him a flute, than to take one who possesses a flute and teach him the art of flute-playing. For nature adds that which is less to that which is greater and more important, and not that which is more valuable and greater to that which is less. Seeing then that such is the better course and seeing also that of what is possible, nature invariably brings about the best, we must conclude that man does not owe his superior intelligence to his hands, but his hands to his superior intelligence. For the most intelligent of animals is the one who would put the most organs to use; and the hand is not to be looked on as one organ but as many; for it is, as it were, an instrument for further instruments. This instrument, therefore, the hand-of all instruments the most variously serviceable, has been given by nature to man, the animal of all animals the most capable of acquiring the most varied handicrafts. (Aristotle, *Parts of Animals*, IV, 10)

This quote has several things wrong with it. For one, Aristotle, the doyen of Greek philosophers, assumed that intelligence was housed in a separate place from the body—some "soul"-ful state—in the mind within and among other elements of the soul such as perception, attention, thinking and so on. Second, it assumes that the body slavishly follows the dictates of the mind. The idea of a mind housed in a brain was not even on the cultural map, at this time. The hands (the body) are simply an instrument of the mind and carry out its commands without any input or feedback. Third, there was no plausible, scientific view of evolution in Aristotle's era. Aristotle advanced the idea of a "great chain of being" or *scala naturae* in which all things were ordered on an ascending scale from least to most complex, with a divine soul at the top of the heap. Hence, the body was almost a secondary accessory of the mind-soul and simply an appendage to the realm of the intellect. This Western tradition has continued to the present day, although there are important alternatives to this view. With this background in mind we begin with an examination of the largest organ of the body, the skin.

The Significance of the Skin

The skin is a fascinating organ comprising over 15 square feet in the typical user and, not surprisingly, the largest organ of the body. Besides its obvious physiological functions as an anatomical barrier to pathogens, temperature regulation, synthesis of Vitamin D, and protection from UV radiation, it also has some very important psychological functions. For one thing, it has remarkable sensory functions that are intimately tied to movement.

> What the organism senses is a function of how it moves, and how it moves is a function of what it senses.[1]

And, in the skin these sensory functions include thermoreceptors for sensing changes in temperature; mechanoreceptors for detecting changes in pressure, texture, and vibration; and chemoreceptors often located on the inner folds of the skin such as the lips, mouth, and inside the nose for identifying tastes and odorants. There are cutaneous nociceptors for sensing pain; some sense toxic chemicals, some sense mechanical deformation of the skin surface, and some sense thermal changes to heat and cold. Polymodal nociceptors, such as the A-delta and C-fibers, relay information to the CNS with regard to chemical stimuli, pressure, and temperature. There are several groups of molecules that underlie nociceptor activity: Peptides (bradykinins), amines (serotonin and histamines), and archindonic acids (prostaglandins). Substance P, released by C-fibers in the CNS, inhibits the perception of pain, whether acute (epicritic) or dull (protopathic). When the nervous system itself is injured it is experienced as neurogenic or neuropathic pain.

All and all, some of these receptors interface with nerve fibers that respond very quickly to temperature and pressure changes while others connect with nerve fibers that respond much more slowly; for instance, what you experience as itching or burning sensations, among others. But skin also conveys our mood (flushing in the face) and physical condition (disease). And, it has aesthetic characteristics that communicate bodily and sexual attractiveness. It also forms the basis of haptic communication and understanding. For instance, touch has therapeutic uses as in massage therapy for back pain or through the use of spinal manipulation in chiropractic work (correction of subluxations or misalignments of the vertebrae). But it's commonly used to convey intimacy (sensual touch), power or both (hand on shoulder) or just simple affection (a hug) or friendship and social conviviality (a handshake).

Even so, there is no credible scientific evidence for the belief in the efficacy of therapeutic touch or the use of the hands to detect and correct the distribution of bioenergetic fields in the body. The claim, in fact, was quietly demolished in an article in the *Journal of the American Medical Association* in an elegant scientific demonstration originally designed by a nine-year-old girl. Her method was simple but ingenious. On one side of a screen stood the therapeutic touch practitioner and on the other side stood the nine-year old girl with holes in the screen to accommodate the practitioner's hands. In order to randomize the order of the presentation of hands of the practitioner, a coined was flipped and the practitioner had to indicate if she could sense the little girl's hands as well as their location in space. Each practitioner participated in multiple trials with either hand. The outcome? Of the 21 therapeutic touch practitioners that participated in the study, their ability to detect an energy field was correct only 44% of the time and below chance expectations, chance of course being 50% given only two possibilities.

An even more strange occurrence is that of fish who possess electroreceptors to detect weak electric fields in their submerged environments. Fish possess several sensory receptors along their lateral line including electroreceptors, chemoreceptors, and thermoreceptors. Electroreceptors are tuned to bioelectrical fields whereas thermoreceptors are sensitive to salinity and water temperature. As it turns out, some fish emit electrical fields detectable by fish and other fish are sensitive to these electrical fields. What is their function? Well, some fish use electroreception to communicate with others in order to stake out territoriality or communicate their position in a social hierarchy among other fish much like primates, including humans. It enables them to find and attract like-minded mates and pick out objects in their marine surroundings such as coral reefs or other inanimate objects. It also helps them identify other kinds of fish and various aquatic animals in the marine environment, some who might not be all that friendly.

Efference Copies and Corollary Discharge

Most significantly it helps them form memories of their environments, particularly significant changes to or deviations from what they have come to expect in their daily aquatic wanderings. In the same way that movement and touch interact to better inform the animal, the movement of fish and other animals impacts on how incoming sensory information is processed. Motor systems in fish and mammals send an efference copy to the sensory systems so that the movement itself is distinguished from action. Another name for this is *corollary discharge* or the ability to distinguish self-generated movement from environmental movement. Did I bump into that rock or is it moving on its own? In this case, it helps meandering fish distinguish between self and other. Between its movement and the movement of other things or neighboring fish. This is known as *sensory-motor coupling* or the ongoing concurrent feedback by its very own sensory systems that better informs movement. In this case, by way of receptors on the surface of the skin. Thus, sensation and movement are inextricably woven.

In human cultures, not just touch, but facial expression, gesture and posture, gaze, spatial behavior among conspecifics or animals of the same species, bodily movements, vocalization, smell, and appearance are all essential and basic to intraspecies communication. Touch, however, is centrally important in the development of the mammalian species, of which humans form a significant part. In one classic experiment, rat pups were stroked with a moist and soft brush when separated from their mother. The effect of the stroking was to counteract the effect of suppression of early growth genes that guide development as would happen if rat pups experienced long-term maternal separation. These guiding growth genes turn on neurochemicals such as growth hormone (GH) and ornithine decarboxylase (ODC), a regulatory enzyme, both of which are essential for normal development.

Shape perception by human infants is also informative. Infants use touch to sense the edges and angles of an object or person to construe its configuration. For instance, an apple is round and has indentures on both sides with one side often having a stem protruding out. And, they use edges and angles along with protrusions of a surface to figure out its unique haptic features. The uniqueness of handles of cups is a good example when they're reaching for their sippy cup.

In addition, they use touch to sense an object's topological structure or how parts of a thing are arranged and interrelated as in using their fingers to explore the contours of their parent's face. Very young infants around 2–4 months begin to perceive shape topologically but not the unique features of objects until about six months of life and not an object's unique configuration until the middle of the second year. Sensitivity to texture doesn't kick in until about 6 to 7 months including the ability to sense the hardness of an object. On the other hand, sensitivity to temperature, size, and volume of an object emerges within in the first few months of birth whereas understanding of the weight of objects develops at the end of the first year.

Circular Reactions, Instrumental Intelligence, and Cross-Modal Perception

Indeed, infants first learn to manipulate small objects through simple clutching in the first few months of life followed by repeated actions on objects in which the infant recurrently examines the object with her hands and eyes. These latter actions according to Jean Piaget, the Swiss psychologist, arise in small stages in the first two years of life. He regarded these as rhythmical and repeated actions on objects or what he aptly named *circular reactions*.

The infant passes through 6 substages of the sensorimotor stage in the first two years of life, according to Piaget. Three of these substages are characterized by the emergence of these circular reactions. The primary circular reaction, around three to five months (substage 2), is body-centered. For instance, the infant continually places her hand in her mouth. The secondary circular reaction at around five to eight months (substage 3) is object-centered. For example, the infant continuously reaches for an object in her crib. The tertiary circular reaction is also object-centered and commences at the beginning of the second year to approximately 18 months (substage 5). In this case, however, she begins to use one object to obtain another or to put it another way, she begins to use two objects in unison to effect an action. For instance, an infant may continually pull on the sheets in her crib to obtain a toy rattle. Piaget called this increasing sophistication in the use of sensorimotor skills *instrumental intelligence* or means-ends knowledge. We might think of this development as the ability to use tools or physical objects to effect an action with the hands. In essence, the infant has begun to physically understand the material relationship between two physical objects and the ability of the hands and mind to influence this relationship with the body. In the subsequent and last substage between 18 and 24 months, Piaget believed that the infant internalizes physical activity as mental activity. He believed that this mental activity was basically thinking with symbols. The problem, however, with this developmental unfolding is that it presumes that after the sensorimotor stage physical activity plays no role in thought and disappears from the intellectual scene, so to speak. I believe, along with a few others, that Piaget was fundamentally wrong on this point and that the body is a constituent of thought, not its mere appendage.

Along with the circular reactions, infants' actions with their hands, on objects, emerges in a parallel set of three phases in the first year. It commences with simple clutching in the first three months to the emergence of rhythmical stereotypies involving repetitive examining behavior during the fourth month. This examining of objects in a repetitive fashion or through the use of behavioral stereotypies in the infant's environment—in extreme, quite common in children with autistic spectrum disorders (ASD)—includes visual regard of their own hands as well as visually-guided reaching for objects in their immediate or proximate environment. By the fifth to ninth months, infants learn to coordinate reaching movements with both of their hands in what are called bimanual activities. This bimanual coordination in manipulating objects with both hands becomes the predominate form of play with toys as well as people, such as "Patti-cake" with parents and siblings.

In all three developmental phases, motor, cognitive, and neurological development interact so that the infant's experience and exercise of their haptic modality is filtered through the background processes of cognitive and motor actions. Likewise, whereas texture, hardness, size, and temperature can be haptically comprehended through the use of the hands and mouth, other qualities such as weight and color are filtered through the visual, auditory and touch modalities.

In humans, it has been demonstrated that sensory systems do not operate independently but rather sensory information is commingled in the brain, a process referred to as cross-modal perception or *synesthesia.* Or to put it another way, there is extensive interaction between the senses. Perceptual illusions, such as the Mueller-Lyer illusion (Figure 2.1), in which individuals are asked to judge the length of identical lines, but in which the context is altered, throws off pretty much everyone even when the subjects are sighted or blind and the objects are visually sighted or physically touched. As it turns out, whether during visual or haptic identification of objects, there is concurrent activation of the lateral occipital area (LO) in what is generally thought of as a part of the visual area of the brain—the visual or striate cortex—not the haptic or somatosensory cortex. Similarly, when an individual is initially primed in one sensory modality and then tested in another sensory modality, both lateral (LO) and middle occipital (MO) areas are activated. In Alzheimer's patients, nonetheless, unconscious or *implicit memory* for objects previously explored haptically is surprisingly preserved. And individuals that lose the ability to recognize faces—a disturbing neurological condition known as *prosopagnosia*—find that they may fail to identify a face not just with their eyes but also with their hands.

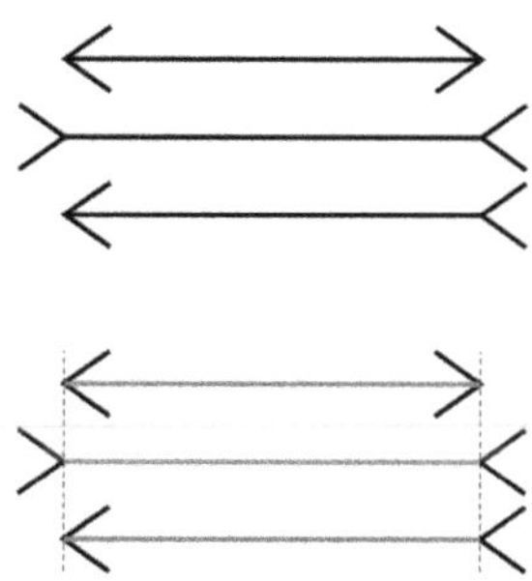

Figure 2.1. Muller-Lyer Illusion.
Credit: Fibonacci (https://commons.wikimedia.org/wiki/File:Müller-Lyer_illusion.svg), "Müller-Lyer illusion", https://creativecommons.org/licenses/by-sa/3.0/legalcode.

The foregoing suggests that tactile perception is widely distributed in the brain in numerous somatosensory regions, motor areas, and visual cortical sites. Can we say anything with more specificity with regard to the neural location of touch? On the one hand, lesions of the anterior part of the postcentral "gyrus" or *ridge* where the somatosensory cortex (SSC) is located—the entire SSC comprises BA1, 2, 3—interfere with various haptic tasks such as the discrimination of cuboidal and ellipsoidal shapes and the ability to distinguish the curvature of spheres. On the other hand, lesions of posterior part of the postcentral gyrus, the intraparietal sulcus, and the supramarginal gyrus impede the haptic recognition of shapes as well as the discrimination of other three-dimensional objects. The intraparietal "sulcus" or *groove* has many functions including the symbolic processing of numbers, visuospatial working memory, visually-guided reaching, object manipulation, and even understanding the intention of others through their natural movements. The supramarginal gyrus (BA 40), one of two main areas of the inferior parietal lobe or lobule, appears to be significantly involved in identifying gestures and postures in others as part of its role in the *mirror neuron system* in the brain. As such, it may be involved in empathic understanding of people but, under stress, which interferes with this ability, may cause us to do the opposite, that is, project our emotions onto others. Empathically, I can understand another by experiencing what it is like to be them and experience what they are experiencing. When this system is not functioning properly, I can't experience their emotions directly but project my own emotions onto them. So, for example, they are not actually angry, but I project my own angry feelings onto them and, as a result, experience them as angry. This may have evolutionary precursors in other animals besides humans as suggested by the following.

Imagine a mother bear with two cubs approached by two humans on a mountain path. With intense fear of the approaching humans she projects onto them an angry threat to her and her cubs and she retaliates with blind rage in an effort to protect her offspring in a manner quite similar to humans. We should remember Charles Darwin's admonition.

> Man with all his noble qualities, with sympathy which feels for the most debased, with benevolence which extends not only to other men but to the humblest living creature, with his god-like intellect which has penetrated into the movements and constitution of the solar system—with all these exalted powers—Man still bears in his bodily frame the indelible stamp of his lowly origin.[2]

Similarly, lesions of the postcentral sulcus in the parietal lobe disrupt the discrimination of the curvature of objects by touch. And there are even additional areas. Haptic object recognition is also processed in the parietal opercular cortex as well as regions in the secondary somatosensory cortices. *Tactile agnosia*, or the inability to recognize objects by touch, is compromised by damage to an important area of the brain, the inferior parietal cortex, which has been implicated in the body's understanding of number such as counting on the fingers. Finger counting, interestingly, is a universal stage in child development and found in all cultures. It has parallels in the historical development of the Arabic number system, which evolved along similar lines as does the development of children's understanding and use of number.

Likewise, tactile form discrimination engages two important but segregated neural pathways in the brain, the *ventral visual pathway* for the processing of vision and the *dorsal visual pathway* that is involved in manipulating tactile form *mental representations* in the mind. That being said, when visual imagery is made use of during haptic perception it may be simply due to the unfamiliarity of the object being explored. Even in the absence of movement, the primary motor, premotor and regions adjacent to, and part of, the supplementary motor areas, are engaged in tactile form discrimination. As a result, some observers have suggested that these areas are constituent of mental processes implicated in decision-making as well as the planning and organization of manual interactions with objects. Let's look next at the role of olfaction in the mind embodied.

Living in the Olfactory Mode

Like most mammals, except humans, licking is part and parcel of normal development. While licking may not be a normal feature of mother-child interaction in humans, bodily contact is an extremely important one. But first, let's examine its progenitor, licking. Felines lick to keep cool, ingest liquids, and to preen themselves, among other things. Snakes sample the air with their tongues to detect prey. Canines show affection and frequently lick their noses in an effort to maintain its moistness for better olfactory perception of smells. It also helps to transfer odorants to the vomeronasal organ in the interior surface of the nose. The vomeronasal organ is situated at the bottom of the nasal cavity. It sends nerve fibers to the accessory olfactory bulb or vomeronasal bulb at the base of the brain. From there, some nerve fibers travel to the amygdala and some find their way to the hypothalamus. What is its purpose? The organ, and its doppelganger in the olfactory part of the brain, detects hormones excreted to the environment by other organisms as well as other odorants, most notably, by conspecifics or other canines, in this case. These chemical odorants are called *pheromones* and are important for communication with like-minded and like-bodied critters.[3] Pheromones, among other things, influence mating behavior in canines and probably similar behaviors in humans although the evidence for their widespread impact in hominins is mixed. Nonetheless, they do impact directly on the hypothalamus, which controls sexual behavior, but they also influence the amygdala, which in canines controls emotional and bodily communication with conspecifics and works similarly in humans and many other mammals. Scent, therefore, clearly and profoundly influences behavior through movement, action, and activity.

The skin of humans also contains many sweat or sebaceous glands, which secrete sebum, an oily substance, and apocrine glands, which secrete androstenes, including pheromones. The latter emerge during puberty. Not surprisingly, apocrine glands are situated on the forehead, the axillary region in the armpits, the areolae surrounding the nipples, the navel, the ano-genital regions, and the areas surrounding the nose and lips. What better places to attract potential mates? But pheromones also have many other important functions. For instance, androstenol, a pheromone, may contribute to menstrual synchrony in women as well as *anxiolytic effects* in humans and other mammals. That is to say, it has a relaxing or anxiety-reducing effect on others because it works as a GABA(A) receptor modulator on brain sites that are affected by this small molecule neurotransmitter, which is inhibitory, and anxiety-reducing. Overall, pheromones and other odors dispersed into the environment have significant effects on individuals and their social behavior.

> Both individual variation and sex differences in perception of these odors suggest that they may play a role in mediating social behavior, and there is now much evidence that they modulate changes in interpersonal perception, and individual mood, behavior, and physiology.[4]

Another structure, the hypothalamus, is located at the base of the brain or *subcortex* and is involved in sexual activity by virtue of its status as the master hormone regulator of the body with areas of the hypothalamus, such as the ventromedial area, concerned with satiety, and the preoptic area, putatively with sexual orientation. The similarly situated amygdala lying deep within the brain in the midline or medial part of the temporal lobes on both the left and right sides of the cortex, hence, bilaterally, is important in emotional behavior, and particularly, emotional learning or the creation and stockpiling of emotional events in memory. In canines, it might consist of a memory of an interaction with a particularly unfriendly dog. In humans, with the remembrance of an unfriendly sales clerk.

Collectively, the *rhinoencephalon* as it is called, consists of the olfactory bulb, the hippocampus, which is involved in the initial encoding of memories, and the amygdala. Evolutionary-wise, it is one of the oldest biological structures in the mammalian brain. Given its unique biological configuration, it is believed to link important elements of smells, memories, and emotions. A smell of a particular perfume or cologne may be linked to an adolescent's first kiss and how it made him or her feel.

Maldevelopment of the hypothalamus, on the other hand, may interfere with the expression of pituitary hormones, particularly testosterone. One medical condition as a result of this misdevelopment is known as Kallman's syndrome. The disorder is marked by anosomia, or the absence of a sense of smell, osteoporosis or fragile and brittle bones, intense emotional fluctuations, and typically, androgynous features or having the physical characteristics of both genders. Brian Brett, who wrote a book about his condition, regarded himself as middlesex, intersex, hermaphroditic or androgyne. He received injections of testosterone throughout his life and described being very hypersensitive to odors. He reached six feet during early adulthood but was underweight and the testosterone injections left him with wide daily fluctuations in his mood state and, he believed, caused him to engage in reckless behavior during adolescence and his early adult years. Whatever may have been the ultimate cause of his reckless behavior, damage to both halves of the brain in the vicinity of the amgydala may help explain why someone like Brian Brett becomes sexually uninhibited and hypersexual.

Adult preferences for certain odors may actually originate in direct experience with specific smells early in life. Odors may encode for a part of the genetic code that is responsible for the biochemistry of the immune system called the major histocompatibility complex (MHC). The immune system, of course, protects us from foreign invaders of our bodies including bacteria, viruses, and fungi such as athlete's foot. The proteins that make up the MHC are found within the membranes of the immune system's cells. Indeed, romantic partners with opposite MHCs are more likely to be sexually attracted to each other. What's the evidence? Studies have shown that women are more attracted to the odor of t-shirts worn by men in the last 24 hours when they both have the opposite MHC complex. However, this phenomenon is attenuated when the women are taking oral contraceptives.

Why are odor and other sensations so important? Marcel Proust, the French novelist living in the late 1800s, wrote seven volumes on the theme of *involuntary memory* in which recollections of the past are evoked without conscious effort by cues from everyday experiences such as odors, musical sounds, particular colors or the taste of a piece of food. But what is so special about involuntary memory? Proust believed that the mnemonic treasures revealed by involuntary memories captured the essence of our individual and past lives.

When I listen to a particular piece of music such as Vladimir Martynov's, *The Beautitudes*, I relive in my mind everything I have ever lost such as my former Yorkie or girlfriend in a continuing chain of past memories. Yet it keeps me tied to that wondrous past or I might forget it entirely. I think it's the minor thirds—or, major sixths, which are inversions of a minor third—in the musical piece that trigger those memories. You may have your own particular triggers. Here's a wonderful passage in Proust from *In Search of Lost Time* in which the mere taste of a small pastry mixed with tea triggers in his mind a flood of past memories.

> No sooner had the warm liquid mixed with the crumbs touched my palate than a shudder ran through me and I stopped, intent upon the extraordinary thing that was happening to me. An exquisite pleasure had invaded my senses, something isolated, detached, with no suggestion of its origin. And at once the vicissitudes of life had become indifferent to me, its disasters innocuous, its brevity illusory—this new sensation having had on me the effect which love has of filling me with a precious essence; or rather this essence was not in me it was me … Whence did it come? What did it mean? How could I seize and apprehend it? … And suddenly the memory revealed itself. The taste was that of the little piece of madeleine which on Sunday mornings at Combray (because on those mornings I did not go out before mass), when I went to say good morning to her in her bedroom, my aunt Léonie used to give me, dipping it first in her own cup of tea or tisane. The sight of the little madeleine had recalled nothing to my mind before I tasted it. And all from my cup of tea.[5]

Cortical and Subcortical Organization of Thought, Movement, and the Body

As introduced earlier, evidence for reciprocal and parallel neural pathways between the so-called motor subcortical structures such as the basal ganglia and the cerebellum as well as areas of the frontal cerebral cortex where it is argued executive functions lie, indicates that there is a very close association between thought and movement. What might this close association be?

Executive functions housed in areas of the frontal lobes in the cerebral cortex consist of such activities as planning and organizing ongoing cognitive activities, monitoring these ongoing cognitive processes, holding them and manipulating them in working memory for ongoing and later use, and delegating attentional resources to the cognitive tasks at hand. For instance, you're getting dressed in the morning for work. Your frontal lobes would be organizing the sequence of activities that makes up your typical morning routine, including monitoring these ongoing activities such as deciding in mid-dress to change one's pants or wear a different shirt. As you go through this typical morning routine, you focus on some activities rather than others and this ability to switch among many tasks (okay, *multitasking*) is part of the role of the frontal lobes in the deployment of one's attentional resources in the various mental tasks one is engaged. And, it's all held in working memory for ongoing use.

These frontal lobe functions are inextricably intertwined with so-called motor areas of the brain. The brain cannot carry out almost any cognitive task without motor involvement. Looking at someone's face for emotional information, reading this text, touching an ancient mole on one's back, orienting one's body to hear a conversation ("pricking one's ear"), swishing a mouthful of wine on one's palette to taste the cornucopia of flavors, scratching one's head in contemplation, playing a divertimento on piano or smelling the aroma of a new sauce on the stove. In each case, movement is intimately bound up with thought. Does scratching your head help you think? We'll get back to this question in the next chapter. Nevertheless, to think is to immanently engage in movement, action and activity.

The neural pathways between the frontal cortex and the subcortical motor structures has suggested to some observers that humans and other species literally manipulate concepts and ideas kinesically, as I mentioned earlier. Is motor activity, then, the basis of intelligence? I presented some arguments for the scientific basis of these ideas in an article I wrote several years ago. It sounded intriguing but given what we know about cognition and about the brain does the scientific evidence even begin to support it? The ability to shift from one activity (doing one's taxes) to another (answering the phone) involves inhibiting the first activity to pursue another course of action or, for that matter, a second, third or fourth activity and results in the production of novel behaviors. We might even think of it as the concatenation of patterned behavior into novel sequences. Of course, answering the phone and switching back to another mental task has circumstances unique to one's own particular case and that is where part of the novelty arises.

Like language, which itself could be thought of as a collection of novel linguistic actions, a finite set of individual actions can be combined into an infinite set of activities. When we speak of novelty or novel behaviors, however, we're not speaking only about the farther reaches of human creativity but about the ordinary, everyday occurrences of novel patterns in behavior. This is so whether speaking or engaging our motor skills such as skiing the North slope of a mountain, adding up items on a bill or figuring out the amount of miles to work, creating musical melodies on a guitar, filming a family event, considering potential ingredients in a culinary dish or discussing a character in a play.

Thus, it has been proposed that there is no inherent distinction between thought and movement at the level of the brain as both can be controlled by identical neural systems. Therefore, concepts and ideas can be manipulated just like body parts in motion. This is because the motor system is actually a complex computational network that controls and directs the brain's circuitry or internal symbols: Counting, timing, sequencing, predicting, planning, correcting, attending, patterning, learning, and adapting. Computation itself is the use of algorithms or rules that act on or process symbols. That is, a set of instructions—such as numerical operations or syntactical rules—that specify what rules operate in what fashion and in what order on a set of symbols—such as numbers, images or words—to accomplish an intellective task such as going over the family finances with your spouse or communicating your disappointment to your boss.

So, behind shifting activities is a discrete combinatorial system that results in an infinite range of potential actions. Language is one well-known example of a combinatorial system. It has the following essential properties. It is discrete, that is, there are individual elements that may be arranged in a potentially infinite range of combinations, and it is compositional, that is, each combination of elements has a different meaning recoverable from the rules or algorithms that inform its composition. But, repetitive stereotyped movement patterns (of any kind)—obsessive-compulsive behavior and Gilles de Tourette syndrome—indicate the malfunctioning of this system. You're literally stuck in a cliched set of actions without recourse from your manual tyranny. Addictive behaviors including gambling, some forms of market investing, drugs, and sex are some common examples, but you also see individuals engaged in stereotypical thinking about politics, about relationships, about how to carry out a particular procedure at work, about what car to purchase, where to live, what to wear, and so on. And the cortex, subcortex, and body work together in communion with the physical and social environment to accomplish these intellective tasks, whether cliqued or novel, frozen or new.

It would be more correct to say, then, that the mind, brain, and body work together in creating intelligent behavior. In fact, the motor system and related structures have three major properties. First, the components of the system are self-organizing, meaning they can organize themselves and evolve through interaction with other parts of the brain and body as well as through interaction with the culture and immediate environment. This is the amazing thing about living systems opposed to mechanical or electrical (computer) systems. Inherent within their parts is the ability to adapt to the environment and organize themselves. Thermostats and computers can't do this. At least not, yet. Second, the motor system is distributed, indicating that the components are part of a larger interacting system, that is, the interaction of mind, brain, body, culture, and environment. Third, these interacting systems are best described overall as *nonlinear dynamical systems*. For one, the system is always changing in response to changes to one of the four smaller, interacting systems. And it's nonlinear because control is not centered in any one part, such as the brain, and the system is organized in a heterarchical manner rather than in a hierarchical fashion. In a hierarchical system, control is located in one system (the brain) and that system controls other internal systems (body) and their interaction with myriad external systems, such as the environment and culture. In a heterarchy, there is true democracy. Each element in the system shares in the overall governance of the system. That is not to say that within a larger heterarchical system, hierarchies do not exist. In much of neuroscience, and science in general, systems are theorized to have a hierarchical level of organization. But nonlinear dynamics posits that much of what's interesting about systems, whether living or nonliving, is that their function and organization can be better understood as a democracy with different parts not necessarily differentiated by their level in a hierarchy but the contribution of each individual element to the overall system. Even the concept of an element theorizes breaking up the world in a particular way to better understand it. Is the nerve cell the fundamental unit of the nervous system or is it just a larger system composed of smaller or larger elements organized in a heterarchical rather than hierarchical fashion? But, a hierarchical analysis of systems is often theoretically useful and easier to understand.

Maybe it's the result of an academic discipline's way of breaking up a problem and studying it from diverse vantage points. Hey, I'm a physicist. I study atomic and subatomic particles. And, I'm a chemist and I research how molecules, that is, larger congeries of atoms and their interactions form the basic building blocks of living and nonliving things. But wait, I'm a biologist and I inquire into the essential nature of cells. And, I'm a neuroscientist and I examine how cells interact to give rise to things like consciousness, attention, thought, and emotion. However, the psychologist hesitates and explains that she studies how culture influences an individual's thought and emotional processes. And, the social psychologist or sociologist looks at how interactions between and among people influences people's beliefs and attitudes, and so on. From the celestial heavens, it appears to be a hierarchy. But these disciplines interact in a multitude of ways and science is only beginning to understand these cross-border connections. So, you could legitimately ask: How are your political beliefs tied to your genes? As it turns out, there is a connection, but it is not well understood. Studies of identical twins suggest that similarities in political beliefs among identical twins are more alike than fraternal or non-identical twins, both groups of which grow up in similar households indicating that our political beliefs can't be all about culture and environment. Authoritarian personalities, for instance, might be inherited. Therefore, only if there were close ties between these disparate discipline's sources of study would there be any reason to think that these kinds of interactions even exist.

What has been proposed so far is that the larger context in which each part is embedded is essential to understanding the overall operation of the system. In the case of the central nervous system (CNS), we might think of it as a massively parallel, adaptive system in which the biophysical makeup of the brain—it's actual physical and biological structure—and its functioning, that is, its neurochemical interactions, are inextricably interwoven. Thus, it is capable of continually updating, modifying, replacing, or generating new neural connections. This is the aforementioned self-organizing principle of living things. Human growth and development, therefore, is not contingent on a central processor, that is, a brain, but emerge piecemeal in development from specific experiences the infant or child encounters through movement and activity softly assembled from his or her interactions with the environment. This suggests a very close correspondence between action, or what we do, and perception, or what we pick up from our sensory receptors about the nature of the environment.

The formation of new neural connections or the process of neurogenesis begins to establish maps or self-generated information encoded in the neural architecture with regard to the relationship among the brain, spinal cord, and motor neurons. These neural maps possess overlapping and *degenerate connections* with input from multisensory sources; principally, audition, gustation, touch, olfaction, proprioception, and vision. This aids immensely in the creation and calibration of perceptual-perceptual, perceptual-motor, and motor-motor relationships. But, a word about definitions here. *Degenerate connections* allow dissimilar brain areas or functions to have similar consequences and these connections affect how perceptual and motor functions are mapped onto each other and get appropriately connected. If you're curious, another kind of connection, *redundant connections*, allow identical brain areas to have the same outcome. Let's examine when these mappings develop abnormally or are out-of-sync.

One instance is ineffective network connections among the dorsolateral prefrontal cortex, thalamic pathways that carry information to and from these higher cognitive centers, and regions of the cerebellum which are involved in timing, rhythm, and other cognitive tasks. These are largely perceptual-motor relationships and can lead to something called *cognitive dysmetria*. Such a condition involves problems in processing, coordinating, and prioritizing information that and can play a significant role in the genesis of psychosis and other mental illnesses. But it can also result in disturbances in language such as *agrammatism* or ineptitude in putting words together, inability to express feelings, deficits in spatial cognition that underlies visuospatial intelligence and spatial memory, and difficulties in planning and organizing everyday tasks, also known as executive functions. This has been termed the *cerebellar cognitive affective syndrome*.

While the cerebellum has been traditionally associated solely with motor functions, there are intimate connections between the cerebellum and the dorsolateral frontal cortex suggesting non-motor or *cognitive functions*. Some scientific observers have proposed that the coeval or joint evolution of the *cerebellum*—particularly, the *dentate nucleus* in the cerebellum that participates in the manipulating and routing of mental symbols such as numbers and words—and the *frontal cortex* as well as the use of the hands and the integration of the hands with *stereoscopic vision* in hominids, demonstrates that motor activity is the evolutionary basis of human intelligence.

The motor basis of concepts and ideas is also suggested from case studies of cerebellar damage where there are deficits in practice-related learning (the wearing of magnifying or reducing prisms that alters visual perception) as well as detection of errors (selecting between synonyms and non-synonyms). These studies demonstrate that the cerebellum may serve as an anatomically distinct long-term memory and learning system or possibly a modifiable *pattern recognition system*. Defects and deterioration of another motor structure, the *basal ganglia*—particularly the substantia nigra, the locus coeruleus, and raphe nuclei—all located within the brainstem, may produce depression or dementia as well as other cognitive deficits. Depression and dementia are both associated with changes to cognition; depression with the intrusion of irrational thought processes ("I'm a worthless person") and dementia with actual neural demise of the brain, particularly memory.

On the other hand, problems with the production of dopamine in the substantia nigra in the basal ganglia may lead to Parkinson's disease in which an abnormal increase in inhibitory signals to the basal ganglia by way of the action of the neurotransmitter, GABA—in the globus pallidus, a major region of the basal ganglia—can lead to a severe reduction in cortical excitation and movement. Thus, the familiar quadrivium of symptoms we described earlier: Bradykinesia or abnormally slow movements, tremor, motor rigidity, and postural abnormalities such as difficulties in standing. In the case of a rare, familial but fatal disease known as Huntington's chorea, there is the opposite effect: An abnormal reduction in inhibitory signals from the globus pallidus leading to a fierce augmentation of cortical potentiation and movement. The symptoms? Uncontrollable dance-like movements known as *chorea* (derived from the Greek meaning "dancing in unison") and rapidly deteriorating cognitive abilities. A gene has been identified so a person can know if they have the disease—it runs in families—but there is no cure. It is caused by an autosomal dominant mutation in the Huntingtin gene on chromosome 4. Autosomal means it's not in a sex-linked gene (X, Y) and because there are 23 pairs if you have just one copy of it, you'll get the disease because it's dominant in character.

With regard to the cerebellum, we know that it plays an important role in timing functions that are utilized by both perceptual and cognitive systems. There is recent evidence, however, that it, along with the basal ganglia, play a role in planning, regulation, and attention. There appears to be two separate pathways for planning movement. The extrapyramidal tracts originating in the brainstem and the corticospinal tracts originating in the motor cortex. The former specifies the goal or target of movement and is a less flexible system. The latter specifies the trajectory of action and has evolved as a more malleable brain system.

It has also been proposed that the cerebellum contributes to basic associative learning processes. The mind largely operates by associating things from similar or disparate realms of experience. Associating colors with words, or faces with names, are prototypical examples. But the cerebellum also has the ability to rapidly shift attention both within and between sensory modalities. This ability to quickly shift attention, unfortunately, may also lie at the core of autism in which maldevelopment of the cerebellum leads to poor social and cognitive development. Autism may occur even in the absence of damage to the amygdala or the hippocampus, which regulate the encoding of emotions or new memories, respectively.

Both the cerebellum and the basal ganglia have similar and corresponding functions. The basal ganglia connect to the dorsolateral prefrontal cortex and stores mental images of spatiotemporal contexts concerned with orientation of the head and body in space. But it also forms linkages to the lateral orbitofrontal cortex influencing our capacity to shift from one mental task to another. As a result, the storage of these mental images or representations in the basal ganglia may assist the frontal cortex in implementing appropriate social behaviors. Forms of social politeness in interacting with others are one example.

A Universal Rhythm of the Body?

I had an interesting experience while I was writing this book. A minimalist fan, I was listening to Steve Reich's "Six Pianos" through my earphones while I was writing on my laptop computer sipping an iced coffee in a café.[6] My right leg (left brain) was twitching in unison to the music. And, then I noticed an infant in a carriage a short distance from me who was kicking her leg at the same tempo. Obviously, the infant couldn't hear what I was listening to. But Reich has often created the complex polyrhythmic structure of his music relying on the base tempo of the human breath. Is there such a thing as a universal rhythm of the body or is this mere coincidence and of no scientific relevance? There is some evidence that consciousness and other aspects of complex brain functioning are created by regular periodic oscillations in the brain, central nervous system (CNS), and the body. Rudolf Llinas and David Pare, neuroscientists, contend that the autorhythmicity of the brain occurs in continuous 40 Hz cycles arising in the amygdala or other cortical or subcortical sites such as the hypothalamus, thalamus, brainstem or entorhinal cortex. However, because there is a 12-millisecond phase shift from the front or rostral part to the back or caudal part of the brain, synchronous events in the brain arise in 80 Hz cycles. This neural humming gives rises to conscious neural states as well as to cognate states such as dreaming in which sensory input is simply detached from ongoing neural activity. Such periodicity also extends to the over 600 skeletal muscles that operate in various aggregated patterns in cycles of 8–12 Hz oscillations per second. These diverse congeries of neurons produce organized movement by way of the olivocerebellar system in the cerebellum and function as movement's pacemaker by way of two major processes: The periodicity of olivary firings and the coupling and uncoupling of olivary neurons, which activate synergies of muscular groups that work in concert to create patterned movement.

Gerald Edelman and Giulio Tononi, also neuroscientists, have suggested that epilepsy in which there are abnormal electrical discharges in the brain is the result of neural states firing in unison due to a reduction in their complexity. Neural states are collections of neurons that fire together forming characteristic brain wave patterns. Sleep is marked by highly synchronized firings of neurons whereas in normal wakefulness there is a continuously regrouping of neural states with changing conditions as is characteristic of normal consciousness. These neural states are constantly changing during normal waking consciousness because what you are thinking, feeling, and perceiving is constantly changing over time. This *stream of consciousness*, as William James proposed in 1890 and who wrote the first psychology text over a hundred and twenty-five years ago, is experienced as a continuous and enduring process.

> Consciousness, then, does not appear to itself as chopped up in bits … it is nothing joined; it flows. A "river" or a "stream" are the metaphors by which it is most naturally described. In talking of it hereafter, let's call it the stream of thought, consciousness, or subjective life.[7]

Attention is part and parcel of this conscious process and has several defining properties. That is, consciousness is always *changing*, is *continuous*, and is *selective*.

> The mind is at every stage a theatre of simultaneous possibilities. Consciousness consists in the comparison of these with each other, the selection of some, and the suppression of the rest by the reinforcing and inhibiting agency of attention.[8]

But, being in an infantile state or listening to music as in my example above, may result from focusing one's attention on a single thing, resulting, like sleep and epilepsy, in a reduction in neural complexity. And the body refracts this neural state through its intimate involvement with the CNS. Or to put it another way, proprioceptive, sensory, and autonomic signals are incorporated in the creation of consciousness as Edelman and Tononi, as well as Antonio Damasio, have separately proposed. In both Edelman & Tononi's and Damasio's formulations nonverbal preconscious and conscious states arise from the mapping of bodily states by various brain systems. These include brain stem nuclei, sensory cortices, hypothalamic and basal forebrain areas, vagal nerves originating in the abdominal areas, as well as the parietal cortex where the brain forms bodily maps of the musculoskeletal system with feedback from our internal milieu and musculature. Cingulate cortices and other midbrain areas—especially the superior colliculi, which collates multisensory information—provide mental representations of the relation of one's bodily self to the environment. The relationship between these preconscious maps and these secondary bodily maps, Professor Damasio calls "feelings." Higher-order conscious states such as self-awareness arise from these preconscious and conscious representations of bodily states and its relation to the environment. Thus, according to these researchers, *consciousness is bodily-based.*

Edelman and Tononi argue that higher-order conscious states are created by the integration of the present with the remembered past. The present consists of perceptual categorization of incoming sensory stimuli. It is as automatic as categorizing an apple sitting in a bowl as a round object that you are about to juggle as well as the remembering of classes of objects that fit into this category of round objects such as baseballs, peaches, and the like. In their formulation, the primitive- or proto-self is the basis of higher-order consciousness. This primitive self encompasses the ability to discriminate among sensory qualia or sensory nuances within a sensory dimension. These nuances include discrimination of different kinds of colors, sounds, tastes, smells, pains, temperature and is carried out by multiple bodily systems comprising sensory, proprioceptive, and autonomic components. The neural framework of these bodily systems is the *body schema* which maps the relation of the bodily parts to each other and to the external environment and thus provides a corporeal context in which sensory discriminations occur. Consciousness, like memory (below), is therefore not simply a neural process. It is thoroughly holistic and embodied.

The integration of this sensory present mediated by the body and the external world with the remembered past stored in memory systems in the brain and body, Edelman and Tononi call the *remembered present.* Thus, they see memory as a form of constructive categorization, that is, memory is the result of the coordination of neural activity by the brain, the body, and the external world and not as an indelible stamp left on the mind as Aristotle set out in his classic text, *De Anima* or *On the Soul.* Memory is not static and representational as Aristotle believed, but is a dynamic and fluid process that is context-dependent, typically ambiguous, and influenced by our prior beliefs and attitudes. Moreover, research has demonstrated that motor behavior is an important aspect of memory. Individuals that use both hands for bimanual coordination—rather than using just one hand in unimanual coordination—when performing a motor task on the Purdue Pegboard Test, had better recall of lists of words on the California Verbal Learning Test (CVLT). In this instance, 16 nouns are read aloud over multiple learning trials and immediate and delayed recall are evaluated. Correlative MRI scans revealed that the cortical surface area in the right lateral prefrontal cortex (rLPFC) was positively related to percentage of correct recalled words indicating the close association between mind, brain, and body.

Proprioception

The proprioceptive sense as coined by the neuroscientist, Charles Sherrington in 1906, is a unique bodily sense and represents a third system or *tertium quid* ensconced between the external or exteroceptive senses (vision, touch, olfaction, gustation, and audition) and the internal or interoceptive senses by which we sense internal changes to our organs including pain. Just like the eyes, inner ear, tongue, nose, and skin, the proprioceptive sense is composed of a number of mechanoreceptors that respond to changes in mechanical pressure and distortion. For instance, some proprioceptive receptors are found in the muscle spindles. They provide information on the length and velocity or speed of the stretching of the skeletal muscles. Other proprioceptive receptors are situated in the golgi organs at the interface between muscles and tendons that attach the musculature to the skeleton. These receptors inform the CNS with regard to the tension of the muscle in its different parts. Another set of proprioceptive receptors is found in the joints and ligaments and provides information to the brain about the relation of the limbs and body parts to each other. While tendons connect muscles to the skeleton, ligaments connect the different parts of the skeleton and set the range of motion of the various joints. Because they are somewhat elastic, they are able to lengthen when under tension and proprioception provides feedback to the CNS on the tension and stretching of these ligaments.

Afferent signals (signals from the body to the brain) from the proprioceptive system are sent by way of the spinal cord to various cortices including the motor, somatosensory, and parietal areas of the brain. In these processing areas, sensorimotor maps of the body are continually updated and revised. They provide a personal, subjective guide to bodily movement and one's interaction with people and objects, including the weight of objects, the position of the limbs in space and time, as well as a map of the external space around one's body that one moves within. If you have ever danced (or, are a professional dancer for that matter) nothing could be more important than the proprioceptive sense to your movement. Avoiding nearby objects with your sense of vision helps, too. In this way, consciousness of one's body provides objective knowledge about physical reality. Rene Descartes, as we discussed earlier, the French mathematician and philosopher of the 17th century, on the other hand, believed that if one could produce a thought then one must exist, at least, in the mental sense: "I think, therefore I am." Sensation and bodily input provided no certainty with regard to the existence of self in Descartes' world. One could argue, however, that proprioception along with our other sensory modalities such as vision, audition, and touch furnish humans and other animals with the neural residue and bodily correlates of the self. They also provide the basis for the fundamental laws of mathematics and physics, and well before these laws were discovered by science beginning in the 15th century, and before Descartes wrote his *Discourse on the Method* 200 years later. To repeat: It might be better said, "I move, therefore I am."

Proprioception has an interesting pedigree in its use in the philosophical study of aesthetics. That is, the idea that an aesthetic sense is wholly grounded in the senses or a single sense. Immanual Kant, the 18th century German philosopher, suggested that aspects of aesthetic judgment were rooted in logical judgments undergirded by feelings of pleasure or desire—as in the tastings on the palate of a wine aficionado—not situated in the individual senses. How could a sensory modality like the proprioceptive sense fulfill the function of providing the cognitive requirements for a logical or cognitive judgment? One possibility is the role played by recently discovered *mirror neurons* in the primate brain. In general, mirror neurons match events observed by a person with actions performed by the very same person. That is, we can match what we do to what we observe and therefore link an action to an observation. There are actually several kinds of mirror neurons, however, and it is not clear what the correlates or connections are, that is, the neural networks that undergird the matching process in human primate brains. Some of these mirror neurons code across sensory modalities such as visual-motor or visual-aural connections; whatever the animal performs, hears or observes. Others code haptic and proprioceptive information or object-oriented action. Touch and proprioception are part of the same somatosensory pathway, which runs from the somatosensory cortex in the brain to the spinal cord and back. It creates the ability of the brain to fashion a representation of the body schema and the basis in humans for their extraordinary imitative capacity: The ability to easily and fluidly imitate actions, gestures, and speech. Imitation is no doubt, the sincerest form of flattery.

Is rhythm related to the evolution of mirror neurons in primates? Rhythm actually doesn't relate directly to proprioception or mirror neurons as it has been localized to a different part of the brain, the posterior part of the superior temporal gyrus. But rhythm is very important in the discussion of a mind embodied as I demonstrated in the origin and expression of consciousness. Nevertheless, besides the intuitive appeal of mirror neurons why should they be important to a proprioceptive sense? Maybe, we could simply imagine what it is like to be on stage or screen and experience what the dancer or actor is experiencing physically. Edmund Jacobson, an American physician, demonstrated in the 1930s that simply imaging a movement, as in generating a mental image of moving a muscle, created electrophysiological activity in the muscle. Proprioception is not strictly necessary. Movement, moreover, including dance movement, is not just about motor action. I will discuss the intimate relation between mental and physical skill in Chapter 9.

Intra-modal correspondences (relations within a sensory modality) and cross-modal correspondences (relations between sensory modalities) in the cerebral cortex probably play an important role in kinesthetic representation; the imaging of the musculature and body in movement. How might an aesthetic sense be informed by such correspondences? For that matter, what are the aesthetic qualities of our bodily movements that the proprioceptive sense might be arousing in us? For instance, does proprioception contribute to the creation and comprehension of form or narrative structure in aesthetic movement? The idea that a proprioceptive sense underlies dance or any other kind of aesthetic perception or production leaves unanswered what, if any role, a sensory modality might be contributing to aesthetic perception. We'll return to the notion of an aesthetic sense underlying the arts in Chapter 7 where I will offer a counterargument to the notion that any singular sense predominantly underlies any particular art form.

Shaun Gallagher, an American philosopher, distinguishes between *proprioceptive awareness* that connotes conscious awareness of one's body and *proprioceptive feedback* or information that is nonconscious and provides feedback to the brain with regard to the position of the limbs and body in space and time and their relation to other body parts and their interaction with objects in the external world. Interestingly, he warns against motor chauvinism or the idea that the only purpose of the CNS is to produce movement, action, and activity. Professor Gallagher does not address, however, the indissoluble relationship between mind, brain, and body that I have argued is constitutive of thought and not merely an accessory.

We discussed some of the non-motor functions of the cerebellum previously and the theory that we use the body to think with. Hiroshi Imamizu, a Japanese neuroscientist, has suggested that the cerebellum houses internal models of objects and tools that has been acquired from the environment. *Mental models* contain causal information about the nature of physical processes and how objects function. For instance, you probably have a mental model of how a wine opener works. In the cerebellum these internal models are believed to reside in the rear posterior superior fissure. There may be additional mental models located in the neocerebellum that assist information-processing centers in the cortex with regard to the use of symbols (words), concepts ("freedom"), and language ("How do I say pass the salt in French?"). Older parts of the cerebellum in the ventral paraflocculus vermis, however, accommodate internal models of the motor system such as mental representations of the positions and velocities of body parts. They also assist in error correction of our neural marching orders with regard to movement of the body and limbs.

There is an interesting phenomenon related to proprioception and metaphorically referred to as *facial vision*. It refers to the sense of pressure on the skin of the face and is often discussed by those who have lost their vision and who must now rely on other sensory modalities to navigate their environment. Indeed, the investigation of this phenomenon commenced with the French philosopher, Denis Diderot, who made several early studies and offered his own prognostications on the subject around the mid-1700s. As it turns out, the *sensorium* of the blind—the totality of sensory modalities available to them—includes sensory impressions of temperature and pressure on the face and bodily surfaces, sound pressure upon the tympanic membranes of the inner ear along with the analysis and discrimination of these very same sonic waves. Proprioceptive feedback from the skin and underlying musculature of the face carefully tuned to the nuances and pressures that the sighted are rarely conscious, gives the blind an added sensory dimension in which to navigate and sample their environment. An interesting instance of sensory exotica rarely discussed in expositions of proprioceptive phenomena.

Biological Motion

The cerebellum is also part of a predictive system that monitors correspondences between what you intend to do and what you actually accomplish. Much of this is based in the motor system. For instance, storing motor commands and their sensory outcomes; sensitivity to the contexts in which movement, action, and activity occurs including the actions to which the body is put; and the tools or objects one uses to carry out a series of actions such as mow the lawn. The brain and body use a special category of motion to accomplish this known as biological motion. Biological motion enables us to infer the internal mental states of others such as what they're thinking and feeling from the movements of human body parts. These include eye gaze; facial expression; gestures of the hands and body; how people carry themselves in their posture, gait, and whole-body movements; as well as the ability to deduce the biological motion of other creatures, such as your pet dog or cat, and what it may represent. We even have a propensity to detect biological motion in non-animate objects as well such as clouds and the fronts of cars—known as *physiognomic perception*—because of our brain hardwiring that has evolved over many hundreds of millions of years.

One cortical structure, the right superior temporal sulcus (STS) allows us to simulate in our mind-brain-body (MBB) other people's actions and associate them to our own motor system. It is a way of using our own bodies as referents for their movements in the detection of their intentions or possible future behavior. The STS is abetted by the mirror neurons we previously mentioned that reside in the ventral premotor cortex in a region known as F5. These mirror neurons provide ongoing dynamic information with regard to our imitation of movements and gestures of others.

Not all of this is conscious, however, as in the frequent use of gesture to support communication through language. Someone, for instance, might point to an object attached to your clothing and you gesture at the same object in almost an identical motor parody in casual conversation. Puce & Perrett, two neuroscientists, propose that mirror neurons provide a *supramodal code for action*. That is, not mental representations or coding of the exact movements themselves, which would be potentially infinite in number, but conceptual models or higher-level action codes and their consequences that humans typically encounter in everyday living. A somewhat crude portrait not a detailed picture of each and every action.

Together with the STS, the right and left inferior parietal areas of the cortex provide a spatial context for the imitation of other's actions (right inferior parietal lobe) together with perceiving the motor parodies of one's own actions by others (left inferior parietal lobe). The right half of the STS appears to perform a visual dissection of the actions of others whereas the left half calibrates the relation between what others do and what we intend to do. Hence, the overall function of the STS and inferior parietal cortices are to differentiate between your own actions and the actions and behaviors of others. These regions together provide a brain and bodily basis for corporeal awareness. In some disordered states such as schizophrenia and other forms of psychosis, there is a failure to distinguish between what one does and what other people do.

Our primate ancestors like infant vervets—small grey, African monkeys—use a motion-oriented categorization scheme to classify objects according to the object's action or behavior. In human offspring, four-year-olds make significant use of motion cues to categorize objects regardless of whether the objects are geometric or animal figures. Seven-year-olds and adults, however, tend to make use of motion cues more often to categorize animal figures alone. Such studies indicate that motion plays a pivotal role in concept acquisition and guides both human and infra-human primates in the categorization of objects and the learning of concepts across the lifespan.

Particularly important here is that the brain tags memory items as category-specific kinds of information. For instance, words for tools, which involve actions, are stored in the left premotor area and the middle temporal gyrus. The names of animals or what you identify visually are stored in the medial occipital lobe. The left premotor area is activated by hand movements that you are able to imagine in your mind and the left middle temporal gyrus by action words that you are able to recall. So, the actional properties of objects including your own body parts are stored in separate areas from objects that you record visually. However, when you store or encode verbal instructions that accompany action, such as verbal instructions that attend a golf swing as in "Pull back both arms as you slightly bend the left knee," both the temporal and parietal cortices on both sides of the brain are activated along with the left middle frontal gyrus. On the other hand when you need to retrieve this information at some later date the anterior cingulate gyrus (the cingulate cortex is the innermost layer of the cortex) together with the cerebellum and areas on both sides of the frontal lobes—the frontal operculum in the lower part of the back of the frontal lobes—are intimately involved along with an area in the parietal lobe, called the precuneus, that is believed to house a sensory map of your body.

Imitation

Imitation of another's movement, action, or activity occurs when one acquires another's behavior by careful observation in which both the goal of the movement and the outward means to obtain it are copied. This is different from most animals where the goal, but not the means, is copied, a term referred to as *emulation*. An observed action that facilitates an innate or learned response, however, is known as *stimulus enhancement*. Some cases of scratching one's head in unison with others is an example of this phenomenon, whereas crowd behavior—clapping together at a performance—is an instance of *response priming*. William Prinz has therefore suggested an ideomotor approach to imitation. When the perception of an event shares features with that event there is an inducement of action. This is because, as we saw with the concept of mirror neurons, the coding of perception and action—what you perceive and what you do—is hypothesized to be undergirded by a common supramodal code in the mind-brain-body (MBB).

In a similar vein, Susan Hurley and Nick Chater, two neuroscientists, have wielded their active intermodal mapping hypothesis. On their view, this supramodal code arises from the comparison of proproceptive feedback from your observation of a behavior with your feedback from your own bodily systems in carrying out that very same behavior or even just imagining it. This feedback is *imitative* when we match what we see with our actions or *intentional* when we motorically parrot what we would like to see in the actions we have observed. We might thus call this *enactive perception*. We understand nonverbally what others do by matching their behaviors to our own internal models, memories, and experience of our own similar behaviors and actions.

The posterior cingulate cortex is most active when you observe someone imitating your own actions whereas the inferior parietal cortex and the superior temporal gyrus (a ridge of cortical tissue situated in the temporal lobe) are most active when you are imitating someone else's actions. And, it appears that the ventral region (bottom part) of the premotor cortex might be where we store sensory images of what we, at least, motorically experience. Andrew Meltzoff and Jean Decety, two developmental cognitive psychologists, have argued that the left inferior parietal lobe calculates the sensorimotor relations that bear on your ability to imitate someone else's actions whereas the right inferior parietal lobe helps you match relevant body parts to both your own imitations and those of others. What is interesting in all of this is that the ability to imitate is even present in very young infants who can match their own body parts (their tongue) to those of others (a parent sticking out a tongue). A distributed neural network including the somatosensory cortex (touch cortex), input to and from the limbic system (emotional subcortex), and other frontal and parietal areas may be central to both the bodily recognition of self and the identification of others including physical objects as we gleaned in an earlier chapter. This bodily network, moreover, also underlies phantom limb sensations. That is, our experience of our physical self, whether actual or real, is created by the same underlying brain and bodily processes. Ronald Melzack, a Canadian psychologist, calls this neural creation a *neuromatrix* and it consists of an amalgamation of visual, somatic, and limbic systems. The synthesis of these neural systems creates a characteristic and unique pattern for each individual called a *neurosignature*. But, this neurosignature and its associated brain and bodily processes is fixed in time—a permanent schema of the body or body schema—not "a continually changing representation of all possible positions of the limbs." This is so because it is predetermined by of our unique genetic and biological heritage for each one of us and modifiable by experience as in a car accident resulting in the loss of a limb.

Imitation is a feature of human learning that is unmatched—no pun intended—in the animal kingdom including cognate abilities found in apes and in Old and New World monkeys. Marco Iacoboni and colleagues at the UCLA School of Medicine argue that the imitation of speech occurs in Broca's area in the inferior frontal cortex. It is here that we match spoken language we hear with the correlative motor behaviors in our verbal articulatory apparatus; the larynx or voice box. Thus, their *direct matching hypothesis* avers that we understand our actions and those of others by matching the actions we observe visually with our motor images or representations of those very actions. When these two systems resonate together, we motorically understand the movement. When the movements are expressive in nature as in a balletic pose or a facial expression, the amygdala (part of the limbic subcortex) and areas that serve the detection of biological motion in the STS are activated. Since there is such a close association between speech and gesture, Broca's area also contains representations of the hands and links the two together as in gesture or signed languages.

For movement, including speech and other actions, a kinesthetic copy informed by proprioceptive feedback from the body is stored in the right superior parietal lobule. We discussed this earlier when we referred to an *efference copy* or *corollary discharge* although historically they have slightly different meanings (Figure 2.2). This has suggested to some that a copy of the motor command that is relayed to the CNS might be what we actually experience as movement or motor imagery and that it involves activation of the same brain areas as the actual physical movement of the body, which is time-delayed. For a split second, we live in the "Matrix."

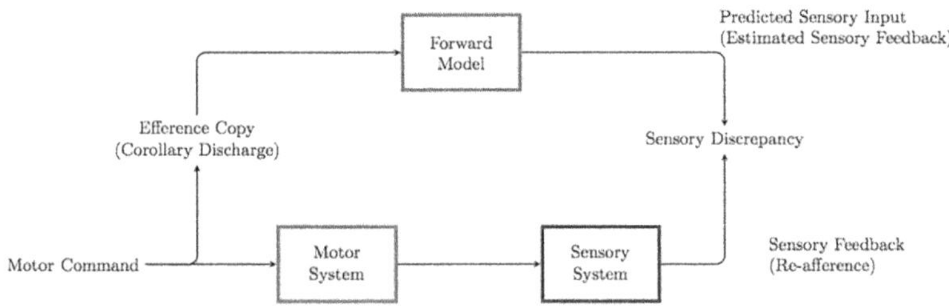

Figure 2.2. Corollary Discharge.
Credit: Gwijde (https://commons.wikimedia.org/wiki/File:Efference_Copy.png), https://creativecommons.org/licenses/by-sa/4.0/legalcode.

Here's how it actually works. The premotor cortex generates an intended movement and the posterior parietal cortex codes the intended movement even if it is not executed. The primary motor cortices—motor cortex, premotor cortex, and cingulate cortices—then capture the neural representation of the motor command, that is, the efference copy. The posterior parietal cortex compares these real or imagined motor behaviors to the efference copy as well as to sensory input from multiple sensory modalities (vision, touch, and audition) and relates it to your unfolding motor plan, that is, your ongoing movement, action or activity.

Motor Control

The planning and execution of movement and the improvement of motor skills with practice are part and parcel of two related aspects; motor control and motor skill learning, respectively. Anatomically distinct areas of the brain work in concert with the body to make these two related components function smoothly together. Motor acts, moreover, can be attended to and hence, *conscious*, or unattended and hence, *unconscious*. The former is subserved by the dorsolateral prefrontal cortex (DLPFC) and assisted by a number of brain structures including the posterior parietal cortex, which integrates vision with proprioception allowing one to coordinate one's own motor behaviors with objects in space and time along with collateral support of the premotor cortex, as in reaching for a glass of water. The dorsolateral prefrontal cortex functions to control the planning of movement in relation to what you want to accomplish motorically and with regard to the locations of objects relative to one another. For instance, finishing the sweater you are knitting so it will be ready for the holidays or executing a billiard shot in which you plan out a sequence of shots such that an initial billiard ball makes several other billiard balls perform as desired including getting the last one firmly into the billiard pocket. Mental practice and learning from observing others performing similar actions can be very helpful in learning new motor behaviors.

The supplementary motor cortex (SMA) along with the assistance of the basal ganglia supports the goal of movement, action or activity in communion with a carefully ordered sequence of acts as in completing a complicated billiard shot. When these sequences are well-rehearsed, they are carried out unconsciously. The red nucleus of the spinal cord (a subcortical, midbrain structure involved in motor coordination) along with the cerebellum—collectively known as the rubrocerebellar system—calculate the minutiae of velocity and force of movement along with the participation of the rest of the spinal cord and musculature. The motor cortex, on the other hand, with the involvement of the supplementary and premotor cortices, sets the spatial parameters of movement. For instance, as you turn your body to grab a glass you are able to avoid hitting any external objects such as a counter top or hot cup of coffee. The function of these motor structures is to code or represent the locations of objects relative to the position of your various body parts. Obviously, this system can go awry as when you and your body are unaware of external objects like a protruding broom handle or a person behind you.

Mark Jeannerod, a French neuroscientist, contends that the ventromedial frontal cortex controls one's responses to the specific properties of objects—their hardness or texture, for instance—whereas the caudate neurons in the basal ganglia help one decide the next step in a sequence of actions as in fielding a baseball or knitting a quilt. As we discussed earlier, patients with damage to the frontal lobes are unable to switch from one task to the next. They may mistakenly pick up the fork just used to eat moo-shoo vegetables, for instance, to stir the coffee. Patients with Parkinson's disease, however, may stay stuck on a particular irrelevant dimension of a previously learned task as in keying into color when the task now calls for use of form such as geometric shape (known as the *Wisconsin Card Sorting Test*). The dorsolateral frontal cortex (DLPFC) calculates movement, action, and activity within a spatial-temporal framework in alliance with the parietal cortex, which gauges extrapersonal space around the body. The motor cortex, however, uses proprioceptive information from the joints and tendons to compute movement. Indeed, there are two major visual pathways in the brain that structure movement, the dorsal or upper pathway and the ventral or lower pathway. The *dorsal pathway* transmits visual information from the visual occipital lobe to the parietal cortex. The *ventral pathway* relays visual information from the visual occipital lobe to the temporal cortex. The occipital lobe is a large area in the back of the brain that processes visual information such as color, form, and movement. The ventral or "what is it? pathway" enables one to consciously identify objects with the body. The dorsal or "where is it? pathway" enables one to unconsciously find objects in space and time. The former helps our bodies with the aid of vision to identify objects and the latter assists our bodies in locating in space and time those very same objects.

Motor commands, as Professor Jeannerod argues, are compared with their resultant effects or what is called a *re-afference copy* of the body's interaction with the environment, such that any differences can be re-computed for more fine-grained and appropriate actions as in learning to hit a tennis ball. Disordered thinking in schizophrenia partly arises from the failure of these very same processes of corollary discharge, that is, a mismatch between the motor command and bodily feedback. The same is true of Gilles de Tourette syndrome, which emanates from a failure to differentiate between competing intentions and the results of any accompanying actions. In schizophrenia, however, one cannot distinguish between self-produced movement and the cognitive intentions behind it, from movement arising from the world outside the body such as the actions of others. In everyday life, however, you understand yourself and gain very useful knowledge about yourself through the effect of your actions on others and from the external environment.

In voluntary movement, as in deciding to raise your arm, the motor cortex (MC), supplementary motor area (SMA), and premotor (PMC) cortices are organized in a *somatotopic* manner, meaning that the body is directly matched, point-for-point, to the corresponding areas in the respective cortices. These point-to-point connections across cortices generate a collection of body maps.

The three cortices (MC, SMA, PMC), moreover, are interconnected in a *homotopic* fashion, meaning that the same relevant parts of each body map relay information to each other in the same area in each cortex. The motor cortex affects the rate and amount of change of muscle force in the various limbs and it receives feedback from receptors in the skin as well as proprioceptive information from these very same muscle groups. The premotor cortex manages initial orientation of the limbs to specific objects by controlling the axial muscles along the center line of the limb as well as the proximal muscles or those muscles closest to the body.

Gender and the Brain

Hormonal influences on the developing brain during gestation and after birth play a powerful role in shaping brain development. The adrenal glands in both genders produce androgens such as testosterone, the major male hormone, as well as estrogen, the major female hormone. Both are known as steroid hormones. There are group differences in cognitive and motor abilities, however. For instance, quick spatiomotor analysis as well as motor coordination, are more often than not found in heterosexual males and less so in homosexual males who tend to be less accurate but better than heterosexual females. Males as a whole are superior as a group on repeated single movements. Males are also better at long-range navigational skills that take advantage of *cognitive maps* or mental layouts of environments. They are also better as a group in the ability to mentally rotate objects as well as dis-embed shapes or objects from their environmental context such as prey in a complex background. Women, however, are superior as a group in employing their fine-motor skills and in performing sequences of movements. Whereas women are more likely to use landmarks as points of reference, men are more likely to use direction, distance, and geometric cues such as the shape of a building. But women may be better at recalling the locations of objects because they may code the location and identity of objects together whereas men may code them separately.

Doreen Kimura, a Canadian neuroscientist, argues that "different cultural groups may have evolved diverse gene pools resulting in different sexual dimorphisms," that is, multifarious gender differences in disparate cultures. Women are more likely to be better at mathematical calculation whereas men are more likely to be superior on tasks requiring mathematical problem-solving and reasoning. Men have putatively better spatial ability in three-dimensional space, but women are speedier in comparing pictorial, numerical, and lexical objects, that is, pictures, numbers, and letters, and tend to possess sharper binocular disparity in perceiving depth. They are also better at interpreting nonverbal cues that occur in posture, facial expression, gesture, as well as in tone of voice or vocal prosody. Women demonstrate enhanced verbal fluency, grammatical usage, and in the spelling of words correctly. Not surprisingly, verbal memory and memory for individual digits are better, too.

Yet, rough-and-tumble play, so ubiquitous in the primate world in males, may be aided by aromatization or the effect of estrogen on the brain as well as the conversion of testosterone to dihydrotestosterone (DHT) and its resulting effects on the amygdala in the human brain. That is, prenatal exposure to androgens in the developing fetus may significantly account for gender differences in preferences for certain kinds of toys, particularly action figures in boys, as well as related activities such as using building blocks or doll play in girls. Variation in exposure to hormones in childhood and adolescence may also be a primary influence underlying sexual orientation in adults in both genders.

Enteric Nervous System of the Gut

There are three major branches of the autonomic or involuntary nervous system of which one, while not as well known, is vastly important to the mind and body. It is called the *enteric nervous system* or second brain (Figure 2.3). The other two are the sympathetic and parasympathetic branches, which regulate functioning of our internal organs. This gut nervous system innervates the esophagus, stomach, and the small and large intestines. It consists of a network of neurons and associated fibers called the myenteric plexus, which is embedded between two layers of muscles that surround the gut. In addition, there is a second neuronal network called the submucosal plexus, which is situated beneath the surface of the lining of the internal mucosa of the gut. The enteric nervous system communicates with the brain by way of the vagus nerve in the parasympathetic branch and the prevertebral ganglia in the sympathetic branch.

Why is the enteric nervous system referred to as our second brain? Just thinking about some mouth-watering morsel of food can cause the parietal cells in the lining of the stomach to secrete gastric or hydrochloric acid for digestion. Indeed, anxiety of any origin can make the stomach more acidic and if gastric acid is returned into the area of the esophagus (throat) it can cause dyspepsia or heartburn, as it is better known. If this occurs over time you may acquire GERD or gastroesophageal reflux disease. Ulcers, too, may be partly caused by anxiety and psychosocial stress even though a bacterium, *Helicobacter pylori*, has been associated with the disease. The standard treatment for ulcers is an antibiotic such as Biaxin or Clarithyromycin along with Prilosec or Omeprazole to reduce the release of stomach acid. But with any anxiety-related disorder psychological treatments to reduce anxiety and stress may also be helpful. Sensory or afferent nerves in the gastrointestinal system send information about toxins or obstructions in the gut and these nerves can initiate the sensation of nausea as well as set in motion the experience of vomiting or cramps. In patients receiving chemotherapy for cancer, for instance, such nauseous feelings can be reduced by 5-HT3 receptor antagonists known as antiemetics, or anti-nausea drugs administered to the gut. This group of drugs blocks serotonin from binding to receptors on the vagus nerve, which is a major neurotransmitter in the enteric nervous system. The vagus nerve and the chemoreceptor trigger zone, the latter which receives inputs from circulating hormones and drugs, communicate with the medulla oblongata, which is the major vomiting center in the brain.

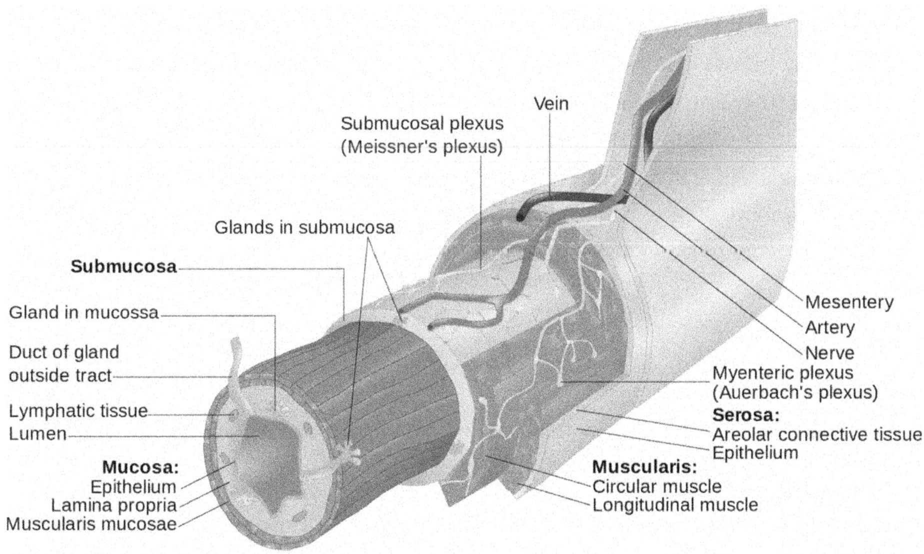

Figure 2.3. Enteric Nervous System.
Credit: Goran tek-en (https://commons.wikimedia.org/wiki/File:Layers_of_the_GI_Tract_english.svg), "Layers of the GI Tract english," https://creativecommons.org/licenses/by-sa/3.0/legalcode.

Thus, there is an elaborate nervous system within the borders of the gastrointestinal system that constitutes the second brain. It communicates with the brain by way of the vagus nerve and the prevertebral ganglia. The prevertebral ganglia innervate regions in the abdomen and pelvis and that convey information about the internal state of the body to the central nervous system. The vagus nerve descends from the medulla oblongata in the brainstem and travels down the neck and shoulders into the abdomen where it innervates the internal organs or viscera. As a result, both the prevertebral ganglia and the vagus nerve interact with the brain conveying internal physical states as well as mood and cognitive states back to the body. Literally, a bodily brain.

Theodore Beauchaine and colleagues at University of Washington have argued that individuals who respond weakly to external reward and also have difficulty modulating their emotions by way of the brain-vagus system, are prone to aggressive and thrill-seeking behaviors or what are commonly referred to as *externalizing disorders*. On the other hand, the opposite pattern ensues among individuals who respond strongly to external reward but who have similar difficulties modulating their emotions. Yet they are typically prone to *internalizing disorders* such as panic attacks and symptoms of anxiety.

Interestingly, there is some limited evidence that the cardiovascular system responds to sensory stimuli such as words, colors, and musical sounds a fraction of a second before the body itself reacts to these sensory stimuli, particularly the greater the emotional importance of the stimulus. Does the autonomic nervous system-vagus-prevertebral ganglia kick in just before the brain actually processes the stimulus? Science currently has no answer to this question.

Although it might not have all the answers, there are some interesting insights coming from very afar—the view from evolution.

Notes

1. Rodney Brooks, formerly a Professor of Robotics at MIT, is now the Chairman and Chief Technology Officer at Rethink Robotics, which is working on collaborative robots for manufacturers.
2. The quote, "Man with all his noble qualities …" is from Charles Darwin, *The Descent of Man*, 1871.
3. The first scientific paper I ever wrote and published was on pheromones in mice when I was in high school during one summer on a scholarship at the Jackson Laboratory in Bar Harbor, Maine: *Implications of pheromones on pregnancy and estrus cycles in female mice*, 1969, *Summer and Academic Year Student Reports*, Paper 1459, http://mouseion.jax.org/strp/1459.

4. The quote, "Both individual variation and sex differences in perception of these odors ..." is from *Current Issues in the Study of Androstenes in Human Chemosignaling*, Havlicek et al., 2010.
5. The quote, "No sooner had the warm liquid ..." is from Remembrance of Things Past (seven volumes, published from 1913–1927) by Marcel Proust.
6. Steve Reich is a minimalist composer who I have been listening to since the early 80s along with Philip Glass and Terry Riley. I met him in the 1980s backstage at Carnegie Hall when I was dating one of his cousins. He is considered one of the greatest living composers of the second half of the 20th century.
7. The quote, "Consciousness, then, does not appear to itself as chopped up in bits ..." is from William James, *Principles of Psychology*, 1890.
8. The quote, "The mind is at every stage a theatre of simultaneous possibilities ..." is from William James, *Principles of Psychology*, 1890.

Bibliography

Andreasen, N. C., Paradiso, S., & O'Leary, D. S. (1998). "Cognitive dysmetria" as an integrative theory of schizophrenia: A dysfunction in cortical-subcortical-cerebellar circuitry? *Schizophrenia Bulletin*, *24*(2), 203–218.

Apostle, H. G. (1981). Aristotle's *On the Soul*. (H. G. Apostle, Trans.). Grinnell, IA: The Peripatetic Press.

Aristotle. (350 B. C. E.). On the parts of animals. (Book IV, Part 10). Retrieved 18 March 2018 from http://classics.mit.edu/Aristotle/parts_animals.2.ii.html.

Beauchaine, T. P., Gatzke-Kopp, L., & Mead, H. K. (2007). Polyvagal theory and developmental psychopathology: Emotion dysregulation and conduct problems from preschool to adolescence. *Biological Psychology*, *74*(2), 174–184.

Bushnell, E. W., & Boudreau, J. P. (1991). The development of haptic perception during infancy. In M. A. Heller & W. Schiff (Eds.), *The psychology of touch* (pp. 139–161). Hillsdale, NJ: Lawrence Erlbaum.

Darwin, C. (1859). *The origin of species*. London: Penguin Books.

Gershon, M. D. (1999). *The second brain: A groundbreaking new understanding of nervous disorders of the stomach and intestine*. New York, NY: HarperCollins.

Hughes, H. C. (1999). *Sensory exotica: A world beyond human experience*. Cambridge, MA: MIT Press.

Ito, M. (1984). *The cerebellum and neural control*. New York, NY: Raven Press.

Ito, M. (1986). Neural systems controlling movement. *Trends in Neurosciences*, 9, 515–518.

Ito, M. (1993). Movement and thought: Identical control mechanisms by the cerebellum. *Trends in Neurosciences*, *16*(11), 448–450.

Jeannerod, M. (1997). *The cognitive neuroscience of action*. New York, NY: Wiley.

Jones, S. (2005). The role of mirror neurons in imitation. In S. Hurley & N. Chater (Eds.), *Perspectives on imitation: From neuroscience to social science*. Vol. 1. Mechanisms of imitation and imitation in animals (pp. 205–209). Cambridge, MA: MIT Press.

Katz, D. (1989). *The world of touch* (Trans. & Ed., L. E. Krueger). Hillsdale, NJ: Lawrence Erlbaum. (Originally published 1925).

Kimura, D. (2000). *Sex and cognition.* Cambridge, MA: MIT Press.

Leiner, H. C., Leiner, A. L., & Dow, R. S. (1986). Does the cerebellum contribute to mental skills? *Behavioral Neuroscience, 100*(4), 443–454.

Leiner, H. C., Leiner, A. L., & Dow, R. S. (1989). Reappraising the cerebellum: What does the hindbrain contribute to the forebrain? *Behavioral Neuroscience, 103*(5), 998–1008.

Leiner, H. C., Leiner, A. L., & Dow, R. S. (1993a). Cognitive and language functions of the human cerebellum. *Trends in Neurosciences, 16*(11), 444–447.

Leiner, H. C., Leiner, A. L., & Dow, R. S. (1993b). The role of the cerebellum in the human brain. *Trends in Neurosciences, 16*(11), 453–454.

Llinás, R. R., & Paré, D. (1991). Of dreaming and wakefulness. *Neuroscience, 44*(3), 521–535.

Marcus, G. (2004). *The birth of the mind: How a tiny number of genes creates the complexities of human thought.* New York, NY: Basic Books.

Melzack, R. (1990). Phantom limbs and the concept of a neuromatrix. *Trends in Neurosciences, 13*(3), 88–92.

Middleton, F. A., & Strick, P. L. (1994). Anatomical evidence for cerebellar and basal ganglia involvement in higher cognitive function. *Science, 266*(5184), 458–461.

Montagu, A. (1986). *Touching: The human significance of the skin.* (3rd ed.). New York, NY: HarperCollins.

Phillips-Silver, J., & Trainor, L. J. (2005). Feeling the beat: Movement influences infant rhythm perception. *Science, 308*(5727), 1430.

Piaget, J. (1952). *The origins of intelligence in children* (2nd ed., M. Cook, Trans.). New York, NY: International Universities Press.

Plomin, R., DeFries, J. C., Knopik, V. S., & Neiderhiser, J. M. (2013). *Behavioral genetics.* (6th ed.). New York, NY: Worth.

Prinz, W. (1987). Ideo-motor action. In H. Heuer & A. F. Sanders (Eds.), *Perspectives on perception and action* (pp. 47–76). Hillsdale, NJ: Lawrence Erlbaum.

Proust, M. (1913–1927). *Remembrance of things past* (*In search of lost time*). (7 vols.). New York, NY: Knopf Doubleday.

Reed, E. S. (1982). An outline of a theory of action systems. *Journal of Motor Behavior, 14*(2), 98–134.

Riedel, M. C., Ray, K. L., Dick, A. S., Sutherland, M. T., Hernandez, Z., Fox, M., Eickhoff, S. B., Fox, P. T., & Laird, A. R. (2015). Meta-analytic connectivity and behavioral parcellation of the human cerebellum. *Neuroimage, 117*, 327–342.

Rock, I. (1985). Perception and knowledge. *Acta Psychologica, 59*, 3–22.

Rosa, L., Rosa, E., Sarner, L., & Barrett, S. (1998). A close look at therapeutic touch. *Journal of the American Medical Association, 279*(13), 1005–1010.

Rozin, P. (1976). The evolution of intelligence and access to the cognitive unconscious. In J. M. Sprague & A. N. Epstein (Eds.), *Progress in psychobiology and physiological psychology* (vol. 6, pp. 245–280). New York, NY: Academic Press.

Schmahmann, J. D., & Sherman, J. C. (1998). The cerebellar cognitive affective syndrome. *Brain, 121*(4), 561–579.

Shima, K., & Tanji, J. (1998). Role for cingulate motor area cells in voluntary movement selection based on reward. *Science, 282*(5392), 1335–1338.

Shimojo, S., & Shams, L. (2001). Sensory modalities are not separate modalities: Plasticity and interactions. *Current Opinion in Neurobiology, 11*(4), 505–509.

Smetacek, V., & Mechsner, F. (2004). Making sense. *Nature, 432*, 21.

Smith, E. E., & Jonides, J. (1999). Storage and executive processes in the frontal lobes. *Science, 283*(5408), 1657–1661.

Soderquist, D. R. (2002). *Sensory processes*. Thousand Oaks, CA: Sage.

Sporns, O., & Edelman, G. M. (1993). Solving Bernstein's problem: A proposal for the development of coordinated movement by selection. *Child Development, 64*(4), 960–981.

Stein, B. E., Laurienti, P. J., Wallace, M. T., & Stanford, T. R. (2002). Multisensory integration. In V. S. Ramachandran (Ed.), *Encyclopedia of the human brain* (vol. 3, pp. 227–241). San Diego, CA: Academic Press.

Stein, B. E., & Meredith, M. A. (1993). *The merging of the senses*. Cambridge, MA: MIT Press.

Stein, B. E., Stanford, T. R., Vaughan, J. W., & Wallace, M. T. (1999). Multisensory integration. In R. A. Wilson & F. C. Keil (Eds.), *The MIT encyclopedia of the cognitive sciences* (pp. 574–575). Cambridge, MA: MIT Press.

Sternberg, E. M. (2001). *The balance within: The science connecting health and emotions*. (Rev. Ed.). NY: Holt, Henry.

Stoddart, D. M. (1990). *The scented ape: The biology and culture of human odour*. Cambridge, NY: Cambridge University Press.

Supa, M., Cotzin, M., & Dallenbach, K. M. (1944). "Facial vision": The perception of obstacles by the blind. *The American Journal of Psychology, 57*, 133–183.

Tecuapetla, F., Jin, X., Lima, S. Q., & Costa, R. M. (2016). Complementary contributions of striatal projection pathways to action initiation and execution. *Cell, 166*(3), 703–715.

Thompson, E., & Varela, F. J. (2001). Radical embodiment: Neural dynamics and consciousness. *Trends in Cognitive Sciences*, 5(10), 418–425.

Tononi, G., Sporns, O., & Edelman, G. M. (1999). Measures of degeneracy and redundancy in biological networks. *PNAS*, 96(6), 3257–3262.

Willingham, D. B. (1998). A neuropsychological theory of motor skill learning. *Psychological Review, 105*(3), 558–584.

· 3 ·

EVOLUTION

Self-generated movement is the foundation of thought and willed action, the underlying mechanism by which the physical and psychological coordinates of the self, come into being. The hand has a special role and status in the organization of movement and in the evolution of human cognition.[1]

Evolutionary Basis of a Mind Embodied

What might be the evolutionary basis of a mind embodied? An evolutionary perspective is important because it sheds light on the adaptive advantages such abilities in humans and other species might serve and for what biological purposes they may have originally evolved. Just like a historical approach is important is understanding current events, an evolutionary approach, not based on an historical time scale but a much longer geological time scale, has important implications for understanding why we became biologically structured the way we are and why we currently possess the mental and physical abilities that we do. The field that studies the evolution of mental abilities and their adaptive advantages is called evolutionary neuroscience as well as the allied fields of evolutionary psychology and evolutionary psychiatry. Charles

Darwin, in the last pages of "The Origin of Species" (1859) speculated on the emergence of such a field that would study the evolutionary basis of mind.

> Psychology will be based on a new foundation, that of the necessary acquirement of each mental power and capacity by gradation. Light will be thrown on the origin of man and his history.[2]

There are three major and many minor parts of our evolutionary story. The dentate nucleus in the cerebellum that has been traditionally associated with movement and motor skills along with the emergence of stereoscopic vision in hominids as well as the astonishing range of abilities that became embodied in the human hand. Moreover, the three became highly interrelated in evolution and together implicate motor activity as the phylogenetic origin or evolutionary basis of intelligence. The dentate nucleus is the most advanced part of the cerebellum that over evolutionary time conspired with areas in the frontal lobes to integrate stereoscopic vision with bimanual use of the hands in primate evolution and push the cognitive envelope forward through movement.

Of the first part, there is evidence of cerebello-thalamocortical loops, a network involving the cerebellum and cortex by way of the thalamus of which the latter acts something like an intricate relay station. This looped network emanates from the dentate part of the cerebellum and travels to the dorsolateral prefrontal cortex, which is involved in spatial working memory, planning, cognitive flexibility, decision-making, and reasoning, and the network as a whole strongly implicates cognitive or non-motor functions. Of the second part, stereoscopic vision enabled humans to gauge depth and distance as well as visually survey a vast 180-degree arc, both vertically and laterally, vastly improving mammalian vision and the capacity for visual thinking and nonverbal thought. Of the third part, as I shall argue, the hands are essential and basic to human intelligence. To paraphrase Vladimir Lenin, the early 20th century social activist and politician, the hand was the first human tool. Or, if you prefer, private property was ultimately derived from labor and it was the labor of the hands that produced and validated private property as one's own: "To mix one's labor with one's hands," to quote the 17th century English philosopher, John Locke.

It has been proposed that thought is really an advanced form of skilled behavior that evolved from earlier modes of flexible adaptation to the environment. Similarly, the body is postulated to be central to mathematical understanding, speech and gesture form parallel computational systems that most likely evolved coevally in language evolution, and mental practice alone has been shown to improve physical skills. On a related note, even now we know that physical exercise can ameliorate or improve some forms of mental illness and might even play a role in improving brain disorder and disease as well as benefiting basic brain health.

One principle of hominin evolution is that the core of human intellect is the capacity of more recent abilities to draw on computational domains that evolved for other tasks. Let's take reading as an example. The capacity did not evolve in humans because the ability emerged very recently in human history. At the turn of the early 20th century the ability to read was not even widespread in the US or other post-industrialized countries. That's because reading is a cultural acquisition; it needs to be acquired and learned and takes years of practice. Yet it depends on other biologically evolved abilities such as visual scanning and the ability to correlate a letter sound (phoneme) with a visual form (grapheme). These latter capacities evolved in humans over many millennia. Writing appeared on the cultural scene about 4,400 years ago in the Fertile Crescent with the beginnings of a formal alphabet following about 600 years later. But writing made use of an area of the cortex that originally evolved for other purposes. This is how culture supervenes on biology and influences the underlying biological architecture as well as the way that architecture functions. In the case of reading and writing, area BA37 in the human cortex is the key area and called the *letterbox area*. This area, Brodmann's Area 37, was named after the neurologist, Karl Brodmann, who investigated the different kinds of cells that make up diverse areas of the brain. In 1909 Brodmann identified 52 separate areas of the brain each with different cell structures. More recently, these differing cell structures throughout the brain have been referred to as the brain's *cytoarchitecture* or regions of the brain with unique structural properties. Since that time some of these areas have been further subdivided but his naming process is still widely used.

In reading, a well-studied cognitive ability, we initially record our visual impressions of letter forms by way of the occipital lobe—actually, the occipital pole at the very back of the cortex—and that visual information is transmitted to the letterbox area in the left ventral occipital-temporal region. It is located on the left side of the brain in the lower half in a region bordering the occipital and temporal cortices. This area it turns out was already prepared by evolution to map things seen or heard to items stored in memory (gestures, sounds, animal tracks), which would have had significant value even in our evolutionary ancestors including other primates and mammals. *De novo* cognitive abilities often arise from capitalizing on brain areas that originally evolved for other purposes. That's how new skills and abilities appear in cultural evolution and presumably change both the brain and the culture in the process. *Bootstrapping* each other using existing cognitive and brain resources to create something new. As the process unfolds in reading and writing, things and events become stored in memory in the temporal cortex where further processing takes place. Things, of course, include stuff like letter sounds, word forms, word and sentence meaning, and the like. They are integrated with sensory experiences such as the sight of an apple or the sound of one's name. Hence, the areas in the temporal cortex are designated *multimodal regions* for processing information from different sensory modalities. They are then assembled into sentences in *convergence zones* in the frontal parts of the temporal cortex known as the temporal pole or temporopolar area (BA38) that assembles items from vision with audition, motion, and meaning. From there they are relayed to the frontal cortices and delivered to unconscious and conscious processes for interpretation.

Here's what happens in the brain when you read a brief paragraph, in this case, a psychoanalytic joke. Two psychoanalysts are walking past each other on the way to work in morning. One says to the other, "How are you doing?" and the other says to herself, "I wonder what she meant by that?"

We call up memories of individual words, their pronunciations, and their meanings from information largely stored in our temporal cortices and we integrate that with our past and current experiences and our current knowledge of the world. We assemble these components in convergence zones that amalgamate individual words and their sounds and visual form, and we process their meanings for understanding in crucial parts of our frontal cortices. Reading didn't evolve by dint of biological processes of evolution but by careful use of areas of the brain and body that evolved for other purposes. Charles Darwin understood that nature is conservative in evolutionary terms. It doesn't need to create new brain or body parts every time an organism encounters a new environment but makes use of existing resources, such as the vertebrate eye or the letterbox area, in harnessing and creating new abilities. When natural selection—biological variations that increase an organism's chances of survival and procreation that are preserved, such as the immense speed in an African cheetah for catching prey—settles on a good "idea" in biological terms molded by mechanical and physical constraints of the real world, it exploits it across the biological spectrum for use in other organisms and contexts.

> Nature is prodigal in variety, but stingy in innovation.[3]

In the third year of life there is an explosion of activity in grammatical language, use of the hands in non-prehensile movements, that is, fine-motor manipulation of objects, and visuospatial thinking as children begin to engineer space in their play with objects and in their budding artistic activities. These developments in human ontogeny were anticipated in primate evolution with a number of singular innovations. Frank Wilson, a neurologist, argues that it began with brachiation or the ability of early primates to swing from their arms. Such evolving abilities in maneuvering between the trees and the forest floor required greater kinesthetic and spatial assets. As the arms and shoulders evolved in concert, it prepared and supported the hands for more complex movements: *Prehensile movements* or the ability to grasp objects, as in an oblique squeeze grip, as well as *non-prehensile movements* or the ability to manipulate objects in a precise manner as in a precision grip. Hand manipulation has an interesting property, too, as it treats the visual, the kinesthetic, and the tactile as one element as the evolving brain increasingly secured the ability to amalgamate the senses as one; known as cross-modal perception.

By the way, raccoons (*Procyon lotor*, the genus designation meaning "dog-like"; the species designation meaning "washer" as they purportedly "clean" or douse their food) have unusually dextrous front paws and evolved from a common ancestor approximately six million years ago. Their paws have five digits just like humans and they are phenomenally sensitive to touch as their cerebral cortex is uncommonly specialized for haptic sensation, maybe more so than any other species outside of *Homo sapiens* (Figure 3.1). They possess whiskers or vibrissae on the tips of their digits and can identify objects literally without touching them, so to speak. They are known to be able to turn handles and open doors, unscrew jars, and uncork bottles and I have observed them on my second- floor outdoor balcony playing in the wee hours of the night. They usually arrive in pairs or packs of twos. They move with all four paws on the railings and through the rafters of the second-floor landing with unusual agility, but carefully. All four paws move in harmonious movement as if choreographed. Unbelievable to watch.

The hand itself went through extensive evolution, too. The opposing thumb of the hand became larger with modifications of the musculature at the base of the thumb. The flexor muscle, in particular, became stronger and more independent, and the metacarpal bones—located between the outer radial bones of the fingers and the inner carpal bones of the wrist—as well as the wrist bones themselves, went through small but significant changes. The fingertip pads became wider and more prominent on the hands. Nine muscles

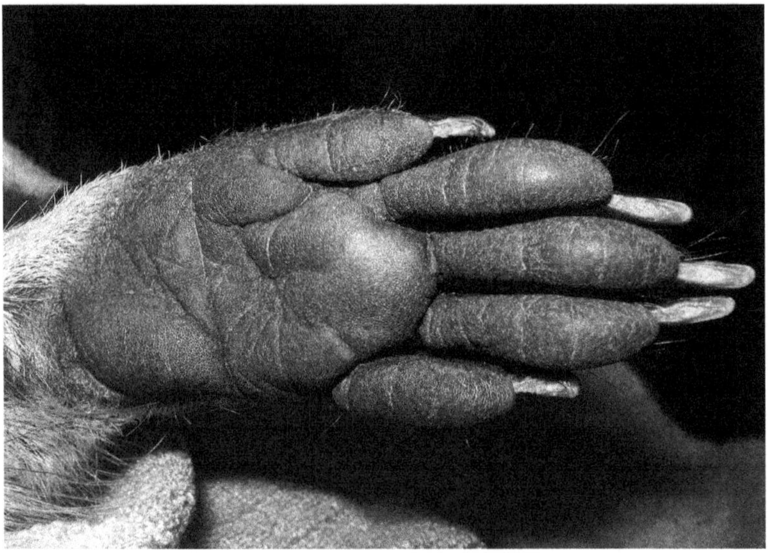

Figure 3.1. Front Paw of a Raccoon.
Credit: Gaby Müller (https://commons.wikimedia.org/wiki/File:Mm_Hand.jpg), "Mm Hand," https://creativecommons.org/licenses/by-sa/3.0/legalcode.

emerged in the thumb for flexion (adduction), extension (abduction), supination (lying flat with the thumb turned up), and pronation (lying flat with the thumb turned down) enabling greater and more intricate fine-motor skills. Handedness came to be dominant in the right side of the brain (known as *brain lateralization*) as the repeated learning and practice of manual skills required a large brain area highly specialized for this process. Dr. Wilson advises that the dominant hand (the right hand in 90% of individuals) is *micrometric* or capable of high precision movements, well-rehearsed, and internally driven, while the nondominant hand is *macrometric* or capable of broad or large movements of the hand, improvisational, and externally driven. Even this tradeoff between the two hands has important evolutionary implications as the two complement each other in interaction with both small and large objects and things. That is, the cultural evolution of tools and tool use (*polylithic* or use of stone tools) on the one hand and the building of larger structures for housing and sustenance (*polypods*) on the other. In our aboriginal ancestors, the development of specialization for tasks among groups of individuals evolved along with symbolic coordination of these tasks, through both verbal and nonverbal communication, social cooperation, the use of collective goals, as well as role complementarity. Of the latter, disparate individuals began to specialize in interdependent roles in the community such as child-rearing or securing food through hunting. They also acquired the ability to assemble objects and things from separately created parts as well as sequence these separate parts in a logical fashion. At the manual level, this is known as *praxis* or the ability to perform skilled movement. When skilled movements become impaired and go awry, they result in a condition known as *apraxia*. Some forms of apraxia are due to loss of the ability to plan movement and others are due to errors in sequencing movement.

> Praxis ... is ... the heterogeneous class of movements that exploit novel biomechanics modifications in the human hand to gain control of external objects.[4]

Stereoscopic vision is another major development in primate cognition along with its precursor, *bipedalism*, or the ability to stand upright and survey the vast horizon of space. Stephen Jay Gould, the former Harvard paleontologist, once called bipedalism, emerging four million years ago in Great Apes, a "catastrophic event." It turns out, however, that the neural substrates of bipedalism evolved much earlier in a common ancestor of *Leucoraja erinacea*, approximately 420 million years ago. This particular living species is a hard-nose small skate, a cartilaginous fish similar in shape to a stingray and related to other sharks and rays, all of which are jawed vertebrates with pairs of fins, scales, and cartilage rather than skeletons of bone. This little skate displays the essential features of gait locomotion in tetrapods—four-limbed creatures, including humans—by way of alternation of the left and right sides of the body and extension-flexion of the pelvic fins. A set of HOX genes regulates the basic overall body plan, head-to-tail axis, and selective innervation of the limb musculature. These genes have been conserved in evolution all the way up to modern humans. In essence, these small skates display walking-like locomotion in a basic neural circuit and body plan still used today by all vertebrates with paired limbs.

It requires an enormous neural network of control processes and musculature to keep a human vertical as well as walk, run, and the myriad abilities that uprightness brings to mind. As I write this, I recall working out in a gym early in the morning a week ago, going for a swim in a hotel pool, and then hiking in the Sonoran Desert with a hydration pack and my Italian hiking boots up a small mountain covered with basaltic-like crumbling rock that made the upscale trek more treacherous than I had anticipated. Only modern humans have acquired so many different kinds of incredibly complex motoric abilities. These include such things as playing the piano or violin, doing a reverse triple flip off a diving board 30 feet above water or from a vaulting horse spinning through the air, putting together a complex watch piece, weaving an intricate sweater by hand, throwing a baseball at 90-miles-per-hour repeatedly for a strike, putting a billiard ball in the pocket in a 3-shot combination play, blowing glass into unbelievably bewildering designs, dancing a balletic piece with 20 dancers on stage for up to 35 minutes with precision timing and rapidly changing complex movement, typing at 160 words-a-minute (my mom claimed she could get up in this range although she could take notes in a mean shorthand, too) not to mention the paintings of the photorealists or the 17-year-old Olympic snowboarding athlete, Chloe Kim, to name just a few. Although many of these movements build on cultural acquisitions (musical instruments, diving boards and pools, finely honed tools, glass furnaces, toe shoes, brushes and acrylic paints, and snowboards and skis), the movements and motions of domesticated or wild animals, as beautiful and complex as they are, often pale by comparison.

Having acquired an upright posture and stereoscopic vision as I mentioned above, enabled humans to gauge distance and depth as well as visually survey a wide 180-degree arc, both vertically and horizontally, significantly improving primate vision and enhancing the capacity for visual thinking or nonverbal thought. Rudolf Arnheim, the first psychologist of art at Harvard, who I met briefly in his later years, argued that *visual thinking* has many singular properties.

> "… is a form of reasoning, in which perceiving and thinking are indivisibly intertwined. A person who paints, writes, composes, dances, I felt compelled to say, thinks with his senses."[5]

> "… 'visual thinking' means that visual perception consists above all in the development of forms, of "perceptual terms," and thereby fulfills the conditions of the intellectual formation of concepts; it has the ability, by means of these forms, to give a valid interpretation of experience."[6]

> "Language, on the other hand, is in itself without form; one cannot think in words since words cannot contain an object. Language is instructed by sensory perception. It codifies the knowledge given through sensory experience … that language is an instrument of that which we have gained through perception, in that it confirms and preserves the concepts."[7]

Visual thinking operates largely through the visual modality whereas language and music, not surprisingly, operate primarily through the auditory modality. For thinking to occur it must be presented with material or data from the senses. The modern human brain and body, the mind-brain-body (MBB) matrix, put their own twist on this data just like any other species or alien (extraterrestrial) being that receives information from the outside world through their own peculiar sensorium. When we're presented with visuospatial information from the environment, we translate and interpret it through our own unique biophysical makeup of body and brain. In our case, we convert biophysical information from light, sound, touch, taste, and smell into electrochemical information, which is subsequently converted multiple times into usable perceptual forms in order to form concepts and ideas. For Arnheim, this visuospatial information is the raw data for the formation of sundry perceptual forms that become the fodder for visuospatial thought. These percepts are the building blocks of visuospatial concepts, which are the intellective tools that we think with, that constitute *visuospatial thinking*.

Rembrandt van Rijn, the 17th century Dutch artist, maintained that he sought to render "the greatest and most natural movement" in his paintings, etchings, and drawings. In his aesthetic explorations in painting, he worked out visuospatial problems in the depiction of light and shadow, contour, motion, space, surface detail, color saturation, use of iconography, and thematic material becoming one of the greatest artists in the history of Western painting. He was, quintessentially, a visual thinker.

Let's consider for a moment a thought experiment that may put the mind-brain-body (MBB) in clearer relief.

A Thought Experiment: A Brain in a Vat

The "Brain in a Vat" philosophical problem is a reminder of how much, in Western philosophy, we have become divorced from our bodies. The problem was formulated by a philosopher decades ago, and its underlying supposition was used as the basis for the movie, *The Matrix*. And frankly, it doesn't make a bit of sense. Nor should it.

Here's the original problem:

> Imagine that a mad scientist takes out your brain and puts it in a vat with life-sustaining liquid. The scientist then connects your brain via wires to a supercomputer, and the supercomputer provides electrical impulses to your brain that are indistinguishable from those that your brain normally receives from your nerves. If the supercomputer could accurately simulate your world, then you would have no idea that nothing you perceive is real.[8]

Let's take apart the argument by looking at the functioning of a simple one-celled organism. Imagine a single cell organism that is completely exposed to the environment including light, sonic waves, tactile sensation, sensitivity to certain chemicals in the environment, and so on. It's a eukaryote so it has a nucleus and can process sensory information gleaned from its environment. It has a specific genetic makeup, evolutionary history, and it has a unique species-specific way of processing sensory information from the outside world and reacting to that information in its own peculiar way including changing its own behavior or initiating new behaviors in response to a stream of constantly changing information in the environment.

If you enucleated the cell and placed the nucleus in a Petri dish with a chemical makeup implausibly identical to its own, it would go own processing sensory information in a roughly similar way as it had before except that its reaction to the environment would have no measurable effect and it would note the discrepancy and react in a new way, note that difference, and continue in an endless cycle until it used up all the available nutrients. In effect, it would cease to be a one-cell organism and just be a purposeless experiment in a dish.

In the human example, without a body, the brain would cease to demonstrate intelligence, its myriad cognitive abilities or anything remotely similar because its own reactions would have no effect on the environment outside of its own internal operations. It couldn't change its behaviors or react in new ways with new thoughts, emotions, or behavior to the outside world. For all practical purposes, it would cease to be human and just be an interesting bit of visual celluloid on screen as in *The Matrix*.

The botheration with the "Brain in the Vat" problem is it assumes that if you take over the sensory or *afferent circuit* then you control the motor or *efferent circuit* and everything that happens in-between. Even more, it assumes that the body plays no role and the brain simply performs a series of inert neurochemical operations without actually participating or reacting to them in any meaningful way in concert with, and through the body; a claim that has no relation to any actual nervous system embedded within a material body.

We know, for example, that corollary discharge is crucial for the MBB to distinguish between self-generated movement and movement in the environment. The sensory system in the body in effect corrects for self-generated movement by sending an efference copy to the sensorimotor cortex. And, such a signal would alert you to the fact that something is out of sync with your own bodily movements or lack of them. Similarly, disordered thought in schizophrenia and other cognitive disorders is the result of impaired internal feedback. Normal internal feedback is crucial to facilitating complex cognitive processes such as problem-solving, attention, memory, creative thought, and the like. Such feedback would alert the MBB that something is amiss. Indeed, in any actual human being, impaired feedback would invariably cause distortions of the body—body-image disorders—because this internal and corrective feedback is crucial to our sense of self and other.

But the real stickler is this. Marcel Proust, the early 20c French author who penned *Remembrance of Things Past* (1913–1927), illustrates through poignant examples involuntary memory or the recollection of "precious fragments" in one's life that occur without conscious effort. Artificial cues from the afferent circuits just wouldn't match up with one's most basic and essential mnemonic experiences. And because memory is an active process, random sensory data pumped into the wet architecture of the brain wouldn't align with the brain's own internal feedback nor with its complex internal history. Each piece of sensory data would also have to be channeled through the neural pathways that align with the five external sensory conduits. Much of what we sense about ourselves comes from within the MBB, however, from internal neurochemical and chemical systems, not from outside. You're tired and fatigued, thirsty or hungry, sleepy, feeling tense or elated, lustful, creative, musical, physical, terpsichorean, and so on. You can't control the experience of any sensing creature from the outside. You can compress it or distort it—sociologist Erving Goffman's total institutions such as prisons, psychiatric wards, military bases, nursing homes, sanatoriums, concentrations camps, boarding schools, servants' quarters, monasteries, and ships at sea—but you can't eliminate it. You can't remove the interaction of the world through the MBB.

> Everything can be taken from a man but one thing: The last of the human freedoms—to choose one's attitude in any given set of circumstances, to choose one's own way.[9]

Nonetheless, it is not just a brain embedded within a skull and within a body, but a *mind-brain-body* (MBB) that ultimately constitutes human thought, intelligence, emotion, motivation, and behavior. You can't simply manufacture these as they are part of what the MBB does as an indissoluble unit in reaction to some alteration in the outside world. It looks for new information and discrepancies and where there is none, that is a discrepancy. A supercomputer could no more anticipate what to feed into an isolated brain that reacts every time it loads sensory information into its afferent circuits than predict the weather accurately, and for similar reasons, because the variables are largely unknown, constantly changing, and potentially infinite. It could never anticipate how the MBB would react and that's the worry because the *Brain in a Vat* problem is a mystification of how the MBB actually operates in interaction with the world and the world in interaction with the MBB, even more so, given that it is actually derived from that very world and a part of it.

Q.E.D.: *The Matrix* is a fiction.

In a similar vein, an article recently published in *Nature*, a scientific journal, claims that thought and emotions might be able to be actually created in isolated brain tissue. Brain tissue in the laboratory would thus have its own *organic vitalism* to produce thoughts, feelings, mnemonic experiences, consciousness, subjective phenomenal states, a sense of self or self-agency or possess sensory qualia, which seems farfetched. This would be true whether the brain tissue has been removed from its human source and re-cultured in the laboratory in some fashion, implanted in a living animal (*chimeras*) or aggregated from pluripotent stem cells (*brain organoids*). However, there is currently no empirical evidence that can decide these scientific questions so true believers can lapse into mere fantasy. But it turns on various inductive fallacies that expose neuroscience's dualism and Jainism vitalism.

Mind, to be sure, doesn't exist in a collection of isolated tissue that has been plucked out of a *mind-brain-body* (MBB) and given a few spritzes of signaling molecules and vascularized to provide some nutrients. Even a previously functioning whole brain can't be sustained outside a material body if such an emergency medical procedure could even be done. The MBB is an indissoluble union rather than a collection of two Cartesian substances–extended (physical) and unextended (mental). The implicit analogy to digital computers is part of the problem.

This latter quandary is a corollary of *The Brain in a Vat* problem. If a mad scientist removes your brain and places it in a liquid-filled vat and hooks it up to a supercomputer that feeds your afferent circuits with putatively identical information is the resulting effect a simulacrum?

But isolated brain tissue, brain organoids, and human-animal chimeras do not, *in-themselves*, have causative powers that give rise to human thoughts, consciousness, feelings, self-agency, dormant memories or sensory qualia. These are emergent properties of the MBB and are not found in individual cells, congeries of cells or larger collections of brain tissues. They are not, in-themselves, imbued with any vitalistic characteristics any more than an assembly of computer chips catalyzes subjective states. As J. S. Mill once opined, "No mere summing up of the separate actions of those elements will ever amount to the action of the living body itself."[10]

Computational Systems, Stereopsis, and the Subcortex

David Marr, in his book, *Vision: A Computational Investigation into the Human Representation and Processing of Visual Information* (1982), published after his early death at 35 years-of-age, proposed three, now four levels of analysis, in describing any computational system that converts raw sensory information into concepts and ideas. We can think of our biology as processing or transducing physical information in the environment to more useful electrochemical information processed by the MBB. Bees use the ultraviolet spectrum to process information that is important to their species but beyond the pale of human biology. Dolphins use sonic waves or biosonar (echolocation). Birds take advantage of the magnetic compass of the earth. Sharks rely on electrochemical detection undersea to ferret out prey. Cats and dogs bet on their sense of smell to lead them to important resources in their natural environments. Ants pick up important chemical aspects of their immediate environment by way of cuticular hydrocarbons on their exoskeleton. There are countless other examples in the natural world.

The first of Marr's levels is the physical implementation of a sensory modality in a biological system such as us hominins or modern humans. In vision, the processing of light by the human eye and the transmission of this biochemical information by way of the optic nerve and the optic pathways is assembled in the visual or striate cortex at the back of the brain creating an initial *primal sketch*. We might think of it as an analogy to an initial pencil sketch by an artist. What algorithms or rules does the brain use to create three-dimensional vision in its wet architecture? This second level is the algorithmic realization of vision, in this case the myriad processing areas of the visual cortex that analyze and output information on color, line, and orientation creating a *2.5-dimensional sketch*. We might conceptualize this as similar to an artist who takes his or her initial sketch and shades in areas of a scene. The third level is the *computational level*. How does the brain assemble shapes, colors, and diverse perspectives—in Arnheim's language, perceptual forms or percepts—to identify objects and things in the real world and imbue them with meaning? That object over there is a pear. That moving shape over there is a giraffe. This is the computational or cognitive creation of a

3-dimensional model where our imaginary artist creates depth using perspective and occlusion of objects to create three-dimensional form. The fourth level is the way culture and environment shape the process of seeing. We call this last process, "learning."

In stereoscopic vision or *stereopsis*, the parallel frontal position of the two eyes in modern humans enables binocular vision, the perception of depth, and the ability to see in three dimensions. Pathologically, there are individuals who lack the capacity for stereoscopic vision and often compensate by using monocular cues for depth rather than binocular ones. Monocular cues include things like occlusion where an object appears farther away because it is partially hidden by an object in front of it. You can detect it with just one eye. Among visual artists there may be even be a higher than expected percentage of so-called stereoblind individuals. It is rumored that Rembrandt himself lacked the capacity for stereopsis and it may have forced him to work out the details of depth and distance in his visual explorations in painting like no other artist that came before him. Like Rembrandt, Leonardo da Vinci appeared to have a visual condition known as *exotopia* in which one eye does not look straight ahead but peers askew and is referred to as a *walleyed eye*. This results in a visual condition known as *strabismus* in which there is a partial or complete inability to maintain alignment of the eyes. As a result, it is believed that there is suppression of the eye that deviates from the object of fixation and two-dimensional monocular vision follows, since binocular vision would result from two eyes visually fixating on the *same* object. But, here's the rub. Intermittent exotopia may have allowed Leonardo—documented in a recent ophthalmological study—and possibly Rembrandt too, to switch easily to monocular vision when covering a canvas with paint thus facilitating the transference of three-dimensional objects to a two-dimensional surface.

Nowadays, one can potentially, at least partially restore stereopsis through practice by way of *perceptual learning* using computer-based technology, among other things. So, for example, research has demonstrated that video game players improve eye-hand coordination, increase speed of reaction times to visual information, significantly improve their ability to mentally image and rotate three-dimensional objects, and augment their skills in attending to two or more things at once (yes, *multitasking*, again). Surprisingly, there is one invertebrate who possesses the capacity for stereopsis or 3-dimensional vision and uses it admirably in prey detection: The praying mantis. To be sure, stereoscopic vision is a complex ability and apparently arose in at least one invertebrate as well as in the Great Apes. When 3-D glasses are outfitted to the tiny praying mantis' two eyes the images distort three-dimensional reality and fool the insect into striking at objects in empty space. Whether the wetware and algorithms that the praying mantis uses to see the world in three dimensions and the brain and body calculations that occur in higher primates are similar, is not known.

Originally, the cerebellum was thought to play a major role in fine-tuning control of the musculature of the body as well as maintaining balance and equilibrium. We have come a long way. It is now known that it plays a role in thought itself. For one thing, there was parallel evolution of both the dentate nucleus, which we mentioned earlier, and regions of the frontal lobe. These two evolutionary developments were integrated with the *coeval* or "roughly at the same time emergence" of bipedalism, stereopsis, and the rapid development and expansion of the dexterity of the hands. Meanwhile, reciprocal and parallel neural pathways appeared between the cerebellum and the frontal cortex where working memory and executive functions such as planning, monitoring, task management, and focusing attention originate. Here's where it gets interesting. It has now been suggested that the evolutionary significance of these pathways is that they enable the *kinesic manipulation of concepts and ideas*. Or, to put it another way, it may be that we literally think with our bodies or, more to the point, with our mind-brain-body (MBB) matrix.

The cerebellum is known to play an important role in timing functions but there is recent evidence that it, along with the basal ganglia, play a role in planning, regulation, and attention. As I discussed earlier, there appears to be two separate pathways for planning movement: The extrapyramidal tracts originating in the brainstem and the corticospinal tracts originating in the motor cortex. The extrapyramidal tracts specify the goal or target location of movement but are less flexible and evolutionarily older. The corticospinal tracts specify the trajectory of movement but are far more flexible. The cerebellum, moreover, contributes to basic associative learning processes—such as associating words with colors—as well as the ability to rapidly shift attention both within and between sensory modalities. It has a significant role in such cognitive activities as the processing of auditory input, verbal memory, language, executive functions, sequencing, and visuospatial perception. And, as it turns out, maldevelopment of the cerebellum may play a role in autistic spectrum disorders (ASD). This constellation of symptoms in autism are categorized under neurodevelopmental disorders because they arise early in life and, in the case of ASD, involve a triumvirate of deficits: Poor social interaction, problems with verbal and nonverbal communication as well as a restricted set of behaviors that are usually repetitive in nature. Constant arm-flapping is one example. This inability to rapidly shift attention in autistic children arising very early in life may lead to poor social and cognitive development even in the absence of damage to the hippocampus and amygdala, which underlie the processing of emotions and help create one's emotional life. But with damage to the *cerebellar vermis*—a structure in the medial and rostral zone of the cerebellum and with connections to all of its major areas—there is emotional blunting and personality changes. The vermis lies in the evolutionarily older paleocerebellum and develops in the first four months of gestation unlike the evolutionarily newer parts of the cerebellum or neocerebellum that develop 1–2 months later in humans. Abnormalities of the vermis, moreover, perpetuate pervasive developmental disorders of which ASD is one, communicative and social impairments, and mental retardation. One theory of autism is that dysfunctional neural circuitry, particularly in the cerebellum (Figure 3.2), results in an inhibitory and excitatory imbalance of neurotransmission that affects both motor and non-motor functions in the autistic individual.

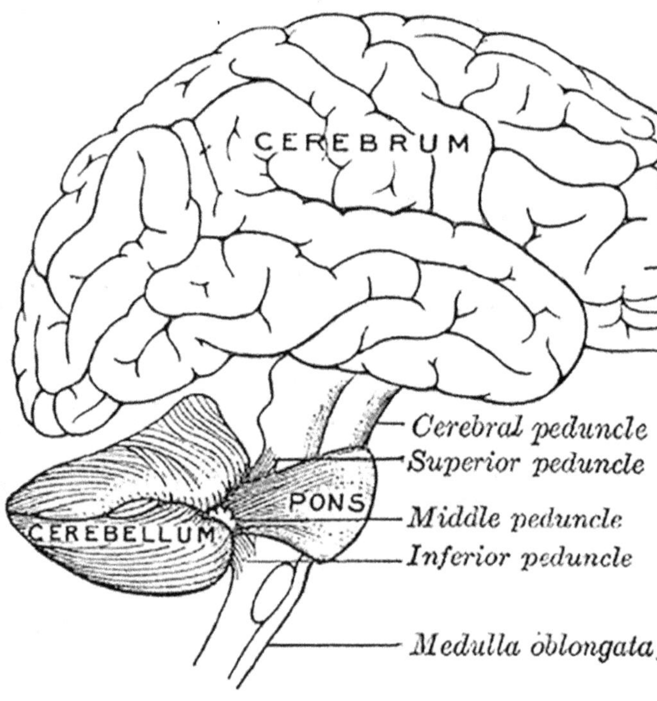

Figure 3.2. Cerebellum.
Credit: Gray, Henry. (1918). Plate 677 in *Gray's Anatomy*. Philadelphia, PA: Lea & Febiger. Public domain.

> The cerebellum is intimately linked to many if not most of the behavioral symptoms seen in ASD. These include disruption or impairments in motor activity and coordination, repetitive behaviors, sensory perception, cognitive ability, and language and speech generation and comprehension. The involvement of the cerebellum in this array of different symptoms is likely a result of the absolute necessity of its powerful and rapid computational power in processing sensory, motor, and cognitive information in the brain.[11]

On the other hand, inadequate network connections among the dorsolateral prefrontal cortex, thalamic pathways, and cerebellar sites leads to *cognitive dysmetria* or problems in the processing, coordination, and prioritizing of information. This dysfunctional *connectome* may play a role in the genesis of schizophrenia and other mental illnesses. Likewise, the *cerebellar cognitive affective syndrome*, mentioned earlier, is linked to disturbances in language (problems in ordering words in sentences or *agrammatism*), personality (blunted affect), spatial cognition (deficits in visuospatial memory), and executive functions (difficulties in planning) as a result of disruption of network connections. The motor basis of concepts and ideas is also suggested in case studies of cerebellar damage where there are deficits in practice-related learning such as learning to navigate your environment using magnifying or reducing prisms and in detection of errors such as selecting between synonyms and non-synonyms. As you can see, the cerebellum has many important functions essential to cognition and thought.

The basal ganglia, by way of connections to, and participation with, the dorsolateral prefrontal cortex, stores information of spatiotemporal contexts concerned with orientation in space. By way of connections to, and participation with, the lateral orbitofrontal cortex, it facilitates the ability to shift from one mental set to another. And the storage of this information in the basal ganglia enables the frontal cortex to carry out socially appropriate behaviors in interaction with others.

Basal ganglia defects and deterioration of the substantia nigra, locus coeruleus, and raphe nuclei in the brainstem upset dopamine pathways and result in cognitive deficits such as dementia and depression as well as bring about motor deficiencies. These deficits and deficiencies are characteristic of both Parkinson's disease in which an increase in inhibitory signals leads to a reduction in cortical excitation and movement and Huntington's disease in which a reduction in inhibitory signals leads to an increase in cortical potentiation and movement. Like the cerebellum, the basal ganglia, a collection of structures in the subcortex or *below the cortex in the brain* are fundamental to thought. Both regions play a significant role in complex cognitive abilities as well as a wide range of motor behaviors, particularly motor abilities that underlie speech generation, hand control and its coordination with visuospatial perception, and disorders like autism as well.

Canines, Chimpanzees, and Corvids

Canines are an interesting case in evolutionary ascendance as they possess many complex cognitive skills and have been enculturated with humans for 40,000 years, more or less. They predate modern humans by 43 million years in early canine-like mammals or *Canids* such as coyotes, wolves, jackals, and dingos as well as later emerging domestic dogs. Modern canines are endowed with a broad arsenal of cognitive abilities. They evince compassion for others, demonstrate strong sensory capabilities such as olfaction and audition, and a basic understanding of causality and its accompanying action. Something like this: "The wind is picking up, I'll head in the other direction to get home." They engage in social play and sociality with others and they can comprehend some sounds of other animals and words of humans as well as read bodily gesture and posture and thus evince the capacity for nonverbal communication (NVC). Canines possess an incipient numeracy as well as a capacity for spatial reasoning and probably an even stronger navigational sense. While their imitative capacities are limited even though some breeds can be trained to imitate gestures and poses, they can emulate the goals of other dogs and humans such as learning the action of sitting from observing the results of these actions in other canines. My Yorkshire Terrier learned to sit when approaching a crosswalk by emulating the behavior of his friend, a toy poodle, who was two years his senior. I never taught him the "sit" command, one reason being that it wasn't a frequent occurrence in his behavioral repertoire. To put it bluntly, he just didn't like to sit unless he had to or because other dogs did it. When he did sit on his own, he would often immediately raise his two front paws with a whining sound indicating that he wanted something. So, his only interest in sitting was a preamble to a request.

Alexandra Horowitz, a cognitive ethologist at Barnard College in New York City, studies the cognitive abilities of canines. In an intriguing experiment examining their capacity for self-awareness, instead of using a mirror self-recognition test that is commonly used to assess higher-order consciousness or self-awareness in human infants, infra-human primates (chimps and bonobos), dolphins, elephants, and magpies and whose primary sensory modality is vision; she used an odor self-recognition test. Canines are notoriously smell-oriented, and they don't typically show a lot of interest in mirrors.

If truth be told, my Yorkie when he was in his first year, reacted to and barked at an image of himself in a floor mirror in a shoe store as if he understood it was himself or another dog, but didn't show much interest in mirrors as he got older—as if he habituated to them or to looking at himself. On the other hand, my sister's Labrador Retriever barks at animals on the television, particularly other dogs. Professor Horowitz found that dogs spend more time investigating their own modified odor than their own natural odor, or the modified odor by itself, indicating that it wasn't the novelty of the odor that interested them alone. Rather, they sense and react to a modified odor of themselves because they possess an incipient understanding of self and other. This suggests that they just might possess some rudimentary sense of self-awareness. My Yorkie would often follow me around the house and listen to my conversations on the phone or with others and barked if I mentioned the names of his canine friends or any significant others known to him. If I went to a certain closet in my home he would wait patiently there and watch me closely to see if I opened the closet door and got out his traveling bag, which meant we would be traveling together and might be visiting his friends (both animal and human) across town. Was he self-aware?

Consciousness itself has a long and deep evolutionary history. In *The Ancient Origins of Consciousness* (2016), Todd Feinberg (neurologist) and John Mallatt (evolutionary zoologist) have demonstrated that perceptual or sensory consciousness, that is, awareness of one's immediate environment, probably stretches back in evolutionary history to the Cambrian Period approximately 520 million years ago. This initial emergence of consciousness in the evolution of animals had two key properties. It supplied animals with insightful awareness of the environment around them, and, in later evolving birds and mammals, it became memory-enhanced. Therefore, animals could now focus their experience on their sense of both the internal workings of their own body and the external world as well as have some memory of what that was like and use that recollection in future encounters. Philosophers call this sense of your surroundings and the objects and things that you interact with *qualia*. It's your subjective feeling about things. Imagine, for a moment, that you are a squirrel and you find that you much prefer these delectable walnuts over here to these dry kernels of corn over there. Don't laugh. I have found that my own backyard brown squirrels have the same subjective experience, or so I imagine from their obvious preferences. They love walnuts, but they are not so keen on dry corn kernels. They generally ignore them unless there is nothing else to eat. So, if they were hungry their body would signal this to them and to which they would become aware and if there was a predator or prey in their midst, they would become aware of that too, and could react or not depending on past experience. This phenomenal experience thus allows the whole of their sensory experience to be nested within their conscious awareness, which is another way of saying that their conscious awareness is a unified experience. When you think of a close friend of yours, you also have a unified experience of them as a whole person—their personality, their laugh, their gait, the way they dress, and their favorite music or movie flics—including memories that you may have had about the times you spent together, but not just as a collection of this or that part of them. This is what sensory consciousness is. But, with this conscious experience also brought with it one of its most important attributes. An animal could now use this conscious experience to influence the world beyond themselves as in reacting to a potential predator or prey or not. If they felt hungry, they could act on the outside world and eat that tasty bug over there and if they weren't hungry or didn't feel like it, they could ignore it altogether. And there were many other things that their sensory consciousness would come in handy in their daily activities, such as where to sleep if they were tired or what tree branch was more supportive. It became essentially a *prediction device* about how to best proceed in their daily lives. Consciousness as a property of nervous systems and bodies thus allowed animals to act on the material world and by acting on it have influence over it. Now that I mention it, when will robots and other forms of machine intelligence have this property that so far has been restricted to the unique properties of biological systems?

Where did this consciousness originate in the ancient nervous system? According to Drs. Feinberg and Mallatt, simple motor reflexes begat sensory consciousness. From an evolutionary standpoint, motor reflexes underlie all movement. At their most basic, motor reflexes are rapid automatic responses to sensory stimuli in the environment. Some are monosynaptic such as a sensory neuron that ignites a motor neuron as in the knee-jerk response that you are familiar with from visiting your family physician as a child. A one-celled organism such as a paramecium in a pond bumps into something and immediately reacts and heads the other way and they have trophic responses to light and chemicals in their aqueous environment and approach or avoid them as need be. These are essentially simple reflex reactions. But some are polysynaptic reflexes such as sensory neurons that send a neural signal to a chain of motor neurons that cause the pupillary light reflex. This reflex controls the diameter of the pupil in the vertebrate eye in response to the intensity of luminescence. Most important, from these simple neural chains evolved the complex circuitry of sensory consciousness. But, two other things were also needed. Basic motor circuits to catapult sensation into subjective feeling or qualia as well as hierarchies of neural control to cement it into place, so to speak. This is where the interneuron comes into to play again. Neural-neural connections, that is, connections among neurons, assemble and organize information from an animal's sensory systems and construct neural maps of its environment that it uses to represent or *image* this very same environment. Or to put it another way, the animal's neural hardware creates a virtual environment of its inner body and outside world to navigate in it as well as record its rovings within this virtual environment. Computers essentially do the same thing but on a more primitive level, which is why it has become so popular to equate what computers do to what humans and other animals might be doing or thinking. When you type on a computer, the computer is, in effect, creating a *virtual environment* that is human-friendly so that you can interact with it. The binary code of the computer would be useless to humans in everyday interaction, so computer scientists have created layers of assembly and higher-order programming languages to construct operating systems as well as application programs such as word-processing software in order for us to harness the computer's usability and, in its most sophisticated incarnations, its potential for artificial intelligence and machine learning. But, in many ways it's still just a fancy digital abacus because we haven't yet figured out how to make it smart in the way humans and other animals are.

Another important feature of sensory consciousness and the evolution of the MBB is the coordinated oscillating patterns of neurons that bind sensory perception. So, for instance, elements of a visual image may be processed in widespread cortical regions but linked together by ɣ-frequency oscillation (ɣ = gamma). These distributed groups of neurons are thus phase-locked and may facilitate the binding of various visual features of an object (color, shape, size, orientation) into a coherent percept or perceptual form for use in identification of a prey or predator as well as visuospatial thinking in the broader sense. Musical sounds would activate another set of phase-locked cortical regions, speech another, and so on.

Related to this are rhythms of the body. *Infradian rhythms* manage reproduction as well as hibernation and migration with cycles longer than one day. *Circadian rhythms* oversee daily biological clocks and *circannual timing systems* predict and anticipate changing internal bodily and external environmental conditions over year-long time periods. Thus, these internal oscillatory patterns of larger neural circuits control sleep-wake cycles and seasonal patterning in mammals and are closely tied to the external environment such as light and temperature. Feeding, for instance, acts as an entrainment signal or *zeitgeber* as does light activating special photoreceptors on the retina of the eye and both assist in modulating the sleep-wake cycles of mammals. But, breeding cycles, humidity, food availability, social contact, and day-length—which of course change throughout the course of the seasons below and above the equator—also affect circadian patterns. Ganglion cells spread out over the surface of the retina send projections by way of the retinohypothalamic tract (RHT) to the suprachiasmatic nuclei (SCN) in the anterior part of the hypothalamus about the luminescence characteristics of light entering the eye. The SCN communicates with the pineal gland, which may play a subsidiary role as it has a somewhat similar evolutionary history as the vertebrate eye, and appears to be involved in photoreception, too. Melatonin, secreted by the pineal gland, plays an important role in the synchronizing or entraining of circadian rhythms that are involved in such things as blood pressure regulation, seasonal reproduction, secretion of hormones throughout the day, body temperature, alertness, and, of course, the timing of human's and other animal's sleep/wake cycles. If you are an adolescent, or are a parent to one, melatonin release from the pineal gland is delayed during the evening and nighttime hours causing adolescents to tend to go to bed later and sleep in during the morning hours. Circadian rhythm disorders commonly occur in shift work where individuals work at night or experience shifts in the hours they work, as the result of jet lag, and in attention-deficit hyperactivity disorder or ADHD, among others. In fact, cognition typically deteriorates when wakefulness is extended into the early morning hours of the next day. A melatonin agonist (in this case, this agonist increases the availability of melatonin in the central nervous system), Tasimelteon, has been approved by the Food and Drug Administration (FDA) for a specific circadian rhythm disorder that affects sleep in blind individuals. In their specific case, because of their blindness, such individuals are unable to maintain a normal sleep/wake cycle and Tasimelteon helps them reset their daily biological clock. Overall, these systems of the body and brain work together to create consciousness, cognition, sleep, and a sense of ourselves, whether hominoid or canine, a bee or a bat.

Getting back to our household friend, the domestic dog may be endowed not just with a sense of self and other, but with the capacity for symbolic communication in their ability to understand sounds as proxies for meaning as in "go fetch the red ball," and to suspend real-world aggression in pretend or symbolic play. Canine communication itself may be symbolic given that dogs convey dominance and submission through gesture, posture, and eye gaze. Canine biological rhythms undoubtedly underlie their sense of time and rhythm, an important feature of living beings as I discussed above. They have the capacity for associative learning and the ability to solve relevant, evolutionarily-prepared problems such as finding available food sources or a mate. Associative learning would consist of the ability to learn by associating things in time (the bell means dinnertime) or contiguity or place (the food bowl is next to the water bowl in the laundry room). In the first instance, a bell is the conditioned stimulus or CS and signals the imminent appearance of food, which is the unconditioned stimulus or US and normally leads to an unconditioned response (eating) or UR. When the bell is heard without the immediate presentation of the food, a conditioned response or CR is created, such as salivation. By the way, if you understood all of this you're at least as smart as a dog.

> Even in the simple form of Pavlovian conditioning, animals show evidence of subtle and interesting cognitive processing: Animals learn about multiple features of the CS and US, the context in which they occur, and the temporal relationship between them. Thus, associative learning is not a low-level process to be contrasted with more cognitive mechanisms.[12]

They also develop sophisticated knowledge of the permanence of objects (Piagetian object permanence, see below) and some basic combinatorial and sequencing abilities suggesting an inchoate grammar without words. Pigeons, too, can learn to comprehend human spoken words and enumerate small quantities as well as categorize social and nonsocial objects such as animals and things, respectively.

Jean Piaget, the mid-20th century Swiss psychologist, argued that children go through six substages in learning the permanence of objects, that is, developing an understanding of the stability and perpetuity of physical entities. So, an 18-month-old human infant in substage 5 recognizes that an object hidden several times in succession from their immediate view still exists and they will crawl or amble to its final location demonstrating a fairly sophisticated understanding of causality. Most adult canines, even when prevented from using odor to detect the object, have no problem in finding the displaced object either. Similar findings have been demonstrated in corvids (crows, jays, and magpies) who, like humans, appear to reach substage 6 in object permanence. Substage 6 is the beginning of *instrumental intelligence*—using two objects in unison to effect an action—in which an animal demonstrates a budding understanding of means-ends relations or tool use. Such an occurrence can occur in a corvid who fashions a stick (means) to pull out grubs from a tree stump (ends) to nourish himself: A stick can thus cause the extraction of food. Even though much younger infants may pass some of these hallmarks before the end of the second year of life, Piaget's general theory of intellective development still appears to be valid. The beginnings of thought originate in movement.

Chimpanzees, however, have the ability to understand *analogy*—comparing two different things as in "I wonder if this coconut shell could function as a cup"—and reflect on their own thoughts or what's known as *metacognition*, as they do when they correct their previous actions or cognitive errors in changing an answer to a memory task designed by a human experimenter. Nonetheless, David Premack, the noted chimpanzee researcher, has argued that if these animals have a grammar it is an *object-action grammar* in which the action and the object of the action are not based on true grammatical categories, but on perceptual categories culled from sensory experience and its accompanying actions. For instance, when a canine picks up a scent and follows its trail, she is demonstrating an inchoate understanding of sequencing and the beginnings of grammar without words. Particularly, as she begins to link these scents together and her subsequent action entails following a scent trail.

Navigational Abilities

The ability to navigate in one's environment is ubiquitous in the animal kingdom. It is indissolubly linked to the mind-brain-body matrix that forms the core of what makes animals intelligent. Mark Hauser, formerly of Harvard, has contended that what is inborn is not the pre-existing knowledge of what is significant to the animal's survival, such as where to specifically find food, but the underlying mechanism for learning. In this case, where the food is. And, this learning mechanism integrates action, the body, and the brain. A *universal toolkit* of which a navigational sense makes up a part consists of the ability to recognize objects that are critical to an animal's survival, such as bananas or conspecifics in chimpanzees, prowess in enumeration or animal arithmetic since quantitative understanding is crucial to procuring food ("I sense there's more food over here than there"), as well as the capability to acquire knowledge about the environment that is critical to a particular species' survival.

The ability to navigate the environment takes many forms in different animals. For instance, honeybees and humans form cognitive or mental maps of the layout of their environments and use them to navigate to food sources or Saturday night's performance of one's favorite opera or play. On the other hand, all animals appear to use *dead reckoning* or the capacity to continually update their speed and distance in reaching some locale based on their past location. Other species use landmarks to plan routes in familiar or unfamiliar environments.

Some animals such as dolphins and bats use *biosonar* (echolocation) or the ability to maneuver and find prey in their environment by using sound waves to locate obstacles and potential quarry. Dolphin sonar is composed of broadband cells in the brain that respond to a wide range of auditory frequencies and reaches well above the frequency range of normal human hearing—which tops out about 20,000 Hz—to 20,000 to 40,000 Hz. Sonic waves in the ocean convey sound energy at about 1376 meters per second, four times faster than sound energy in the air which ripples through the ether at about 344 meters per second. Bat sonar works in a similar range and, as it turns out, certain moths that bats feed on have evolved the capacity to engage in evasive actions when bat sonar targets them. How does bat sonar work? When a bat approaches a target, let's say an edible moth, changes in the velocity of approach of the bat (the rate of change of her position) are represented in neural maps in the brain that change their firing pattern as the velocity changes. This is known as *computational anatomy*.

And then there are biological compasses. Besides cognitive maps, many birds make use of both sun and magnetic compasses to navigate their aerial or arboreal habitats. For instance, birds can make use of the position of the sun yoked to the time of day to fly in a certain direction. That is, they acquire a celestial reference point and use it to navigate. On the other hand, honeybees rely on their sensitivity to ultraviolet light to navigate using the position of the sun. What appears to be involved is that the celestial pattern of polarization of sunlight is fundamentally the same as the pattern of polarization of the receptors in the bee's eyes. The bees simply match these two and the bee is off and flying to the appropriate food source or distant hive. The computational anatomy is embedded within the bee's optic system. In desert locusts, *Schistocerca gregaria*, the integration of the intensity of polarized light reaching the brain and chromatic cues to locate the sun's position are housed in the neurons of the *central complex*. This neural structure is in the center of the insect's brain and consists of the central body and the protocerebral bridge. It is sensitive to polarized light from the sun and is topographically ordered to the layout of insect's body so that it serves, in essence, as a compass-mediated bodily navigational device critical to the insect's survival. It is central to the most basic of locomotor behaviors including walking and flying. Single-cell recordings from the central complex show that the insect brain processes self-generated movement and the two together fashion locomotor activities in desert locusts. Many birds, however, make use of the angle of the magnetic field of the earth to navigate large distances, a process called *magnetoreception*. In the case of birds, but not humans, the pineal bodies in the brain contain ferro-magnetoreceptors that respond directly to sunlight and it is able to pass through their thin skulls and, as a result, provides them with two sources of navigational information, both solar and magnetic counsel in their travels.

Electroreception in fish is another navigational technique unique to certain species of fish. Along the fish's body at about the midpoint are embedded electroreceptors, chemoreceptors, and thermoreceptors. Why it is not completely clear, these electroreceptors may respond to electrical fields generated by the ocean's currents providing a biological compass for fish to navigate. Some of these electroreceptors also sense the electric fields of other fish and can be used for social communication such as the formation of social hierarchies, to locate and secure mates, and establish territorial dominance. Indeed, the fish's movements themselves play a special role by modifying incoming sensory information including input from touch receptors as I mentioned earlier. The sensory feedback from locomoting fish refashions motor activity and this complimentary process provides a more robust source of information than mere sensory input alone. As we previously discussed, efference copies of electrical discharges—in essence, neural surrogates for movement—are fed back to the sensory system to negate the movement results of the electroreceptors themselves that impact on these sensory systems. Movement is thus constantly modifying incoming afferent (sensory) input as the two are in indissolvable union.

Pheromones or "ecto-hormones" are used by some creatures to attract or repel individual members of the same species. Pheromones are hormones secreted outside the body to signal conspecifics. There are alarm pheromones that communicate danger or fear, dispersion pheromones that signal one to keep at an appropriate distance and used to establish territoriality, and aggregation pheromones that signal food sources or optimal nesting sites, among other things. In mammals, pheromones are detected in the nasal cavities by way of the vomeronasal organ I mentioned earlier. The organ has connections to the accessory olfactory bulb at the base of the brain that projects to the hypothalamus. Areas of the hypothalamus, such as the preoptic area, play a role in sexual behavior. In humans, as we discussed in Chapter 2, apocrine glands in the skin secrete androstenes beginning in puberty. They are located on many areas of the body including around the nose and lips, on the forehead, in the armpits or axillary region, surrounding the nipples of the breasts, and the ano-genital and naval regions. Sebaceous glands in the skin that secrete sebum also contain pheromones. Thus, pheromones may play a significant role in attracting mates in humans, among other things.

Ludic Relations (Play) in Animals

The idea of play in animals is controversial. Dogs and cats, of course, may chase their tails as if engaged in play but is it really play? Play is typically defined in humans as pleasurable activity that is not goal-oriented as well as involving active engagement that is voluntary and spontaneous. Play in animals, however, sounds eerily similar to play in humans. Canines initiate play through the use of a play bow that signals to the other canines that the initiating animal wants to engage in ludic relations. One way that the play bow might engender the mood for play in the recipient canine is through a *kinesthetic mapping* of the initiating animal's body movements onto the recipient's own bodily movements. As Nicholas Humphreys, the British social scientist, has suggested, such kinesthetic and emotional mapping—the same feeling state is imputed onto the recipient animal—may be the way by which canines and other animals, including humans, treat conspecifics as alternative versions of themselves and thus attribute to them the same intentions, knowledge or affective states as their own. Mark Bekoff, a cognitive ethologist, contends that canines may switch dominant and submissive roles in play as occurs when a typically dominant dog rolls on his back when engaged in harmless play-fighting. Within these play bouts, nonverbal communication (NVC) through gesture, posture, and eye gaze may be a way that dogs share information about themselves with each other. Indeed, canines will also take part in self-handicapping to extend the play interaction. In these cases, they will ease up on the aggressiveness of their play bouts such as damping down their mouthing actions with other dogs.

Animals typically draw on behaviors used in other contexts (predation or aggression) in the activity of play, but that now have acquired a completely different meaning given the kinesthetic and emotional context of play including play with others or so-called social play. Social play in canines probably has some constituent of its origins in that fact that domestic dogs have been enculturated by humans and have lived in close in interaction with us for 40,000 years or more.

Communication

Animals communicate through sounds, postures, gestures, and facial movements such as eye gaze. Animals certainly communicate using a variety of means. But, does their communication constitute a language? Charles Hockett, the American linguist, argued that any true language must contain certain design principles including the ability to both comprehend and produce the sounds of conspecifics (*interchangeability*), the means to refer to other conspecifics, objects or events that are non-present (*displacement*), the power to convey meaning through communication in a one-to-one correspondence of signal to referent (*semanticity*), and the proficiency to combine communicative signals to create new meanings (*productivity*). Canines, for instance, seem somewhat capable of the first three organizing principles of language, but not the fourth, a fact only occasionally found outside the primate kingdom and only fully implemented in modern humans. Nonetheless, dogs can replicate each other's sound patterns as well as respond to the pitch, frequency, and temporal dynamics of other canines and can even recognize, in some individual cases, 100–200 human speech sounds, including individual words. Some observers of canine communication argue that dogs put together in logical order both communicative nonverbal gestures as well as vocal signals suggesting an inchoate grammar. And, although they may have a grammar of the body as I discussed earlier, they may also have an embryonic grammar that applies to communicative sound. Grammar, as defined by contemporary linguists, is a discrete combinatorial system. Such a system has the following properties. It is composed of discrete elements that can be arranged in potentially infinite combinations and each of the resulting combinations gives rise to unique meanings that are the consequence of the principles and rules of language that operate on its distinct parts. In humans, it is hypothesized that there is a mental parser that evaluates sentence structure and returns an initial analysis during language comprehension—encoding words into thoughts—according to various rules of grammar. For instance, some of these rules have historically been visualized as phrase structure trees. A noun phrase (NP) may consist of a noun (N) and a prepositional phrase (PP) or NP = N + PP as occurs in the following sentence: Steven (N) went to the store (PP). More recently, a simple *Merge command* has been proposed to explain the grammatical basis of human languages. Take two syntactic units (read, books), recombine them as one ("Susan reads books"), and recursively combine them again to create hierarchically structured sentences ("Susan reads books that make her smart"). Interestingly, the supposition is that "human speech recruits evolutionarily ancient motor learning circuits," which is a truism if speech were to evolve in humans in the way that I believe it did. These linguistic rules or algorithms are theorized to be what the human mind uses to understand and produce comprehensible speech along with a mental dictionary or *mental lexicon* of learned words and phrases, among other things. These other things include a vocal tract, the ability to perceive speech sounds (phonemes), phonological rules (how sounds are combined), morphology (how words are combined), as well as map phonemes onto written words (graphemes).

Clearly, canines lack some of these capacities because they haven't fully evolved the biological structures that underlie these mental and physical abilities. They can only perceive a limited class of constituent sounds and they appear to only be able to link single words to identifiable objects, although these may be quickly learned in a single exposure or two. This is referred to as *fast-mapping* or the mapping of a sound to a visually perceived object in the mind. Some have suggested that the mapping is not so fast but rather a slow-mapping but, at least in border collies, that is not true. For fast-mapping to occur canines would have to understand that objects have labels, possess the ability to learn by exclusion, and have the mental capacity to associate these word-object associations in memory. A border collie, "Rico," learned the labels of over 200 different objects and was able to infer the names of novel objects by exclusion as well as retrieve them four weeks later. This suggests that fast-mapping is a general learning and memory mechanism and not directly linked to the extensive human-like language network or, at least, not to the syntactic aspects of human speech.

Such specific skills, however, appear to be absent in chimpanzees. Vervet monkeys use alarm calls to signal predators and appear to use communicative rules to combine these various alarm calls but they do not possess the ability to recombine these calls in new ways to mark new types of predators or fresh concerns about their environment. On the other hand, bonobos—closely related to chimpanzees—of the Great Ape family, which are the closest evolutionary species to humans, appear to possess rules for ordering signs. Some of these types of signs pick out abstract categories such as gender suggesting the possibility of an intermediary stage in the evolution of language. Even so, dogs must have some kind of primitive lexical dictionary and the ability to understand simple phrases such as "lie down" or "go get in your bag" as my Yorkshire Terrier was fond to do. The question is whether they process the phrases as constituents of discrete elements <go + get + in + your + bag>, isolated signs <bag> or whole phrases <go-get-in-your-bag>. As the canine ethologist, Marc Bekoff, notes, "dogs are dog-smart" and in the world of canines, sounds, barks, and growls have their own unique language. Whether they are combined as meaningful elements or simply repeated endlessly in a compulsive fashion is an open question particularly when my Yorkie is hankering for a bone (in his world, a "bone" is a dog biscuit). Thus, intelligence and communication in canines and other animals means intelligence and communication for what evolved purpose? Communication has evolved in each animal's habitat and environment for some purpose that has ecological relevance to that species. If not, it has not been selected for in evolution.

Michael Corballis, a cognitive scientist, argues that spoken language likely evolved from facial and manual gestures. In his early theory of the origins of spoken language, Professor Corballis proposed that a system of gestures of the arms, face, hands, and mouth evolved first, followed by grunt-like sounds accompanying these very same bodily gestures. Subsequently, *invisible gestures* or movements of the oral cavity, ushered in a syntactical gestural language something on the order of 500,000 years ago. Modern click languages practiced in sub-Saharan Africa may indicate a form of language that predated full vocalization. From that point, spoken language evolved, possibly as a result of the cultural evolution of language due to indirect actions, such as the *Baldwin effect*, that is, learned behaviors are slowly incorporated into the genome, rather than the direct effects of biology. In mature languages, manual and vocal components began to serve different but complementary uses.

However, the significance of this evolutionary sequence is not entirely clear. The capacity for serial order and an alleged proto-grammar in infrahuman animals suggests that core components of speech may have emerged from the ability to engage in behavioral sequences that were later mapped semantically onto objects that could then be recombined nonverbally in some limited ways. Following this, the capacity for recursion arose enabling the embedding of object-action sequences within one another. Thus, a rudimentary syntax of object-action relations was established and constituted the foundation of a nonverbal protolanguage with some embryonic capacity for combination and recursion. Recursion in formal linguistics means the embedding of phrases within phrases but in the context of a nonverbal *proto-grammar* it would mean the embedding of an object-action sequence within another object-action sequence. For example, <bark + move to location of "owner"> is embedded within <go to kitchen + bark> in a request for food. That is, the object-action relationship is returned or rerun within the sequence of one of its steps as in the case of my 7-year-old Yorkshire Terrier, *Canis familiaris*.

Yet, bottlenose dolphins, *Tursiops truncatus*, can distinguish gesture-action order and recombine those grammatical strings into meaningful patterns that reference concrete objects. For instance, in the sequence of mimed gestures <surfboard + person + fetch> the dolphin will correctly carry out the object-agent-object relationship by taking the person to the surfboard. However, when the sequence of mimed gestures is recombined into a different sequence, <person + surfboard + fetch>, the dolphin will experience no difficulty in doing the opposite and taking the surfboard to the person. Bottlenose dolphins, moreover, comprehend symbolic references to absent objects (displacement); correctly move to an object when pointed to putatively unlike some chimpanzees in the wild; continually update working memory by incorporating previously performed actions; attach gestural symbols to named body parts (semanticity) and perform actions on body parts with or without the use of objects as in touching an object with the named body part; imitate actions performed by other dolphins or humans, both directly and vicariously (TV monitor); and engage in vocal, not just behavioral, mimicry (interchangeability). In the case of behavioral mimicry, bottlenose dolphins appear to use their body plan as an analogy to the body plan of humans. For instance, using their tail as a concrete analogy to the leg. The question is whether bottlenose dolphins or other species outside hominins can amalgamate or merge communicative signals in novel ways to create new meaning using Charles Hockett's scheme of productivity. It appears that they can comprehend new orderings <person + surfboard + fetch> but can they creatively produce new arrangements of syntactic objects on their own?

The dolphin brain has many similarities to hominids including the spindle neurons in the neocortex. The neocortex of many species is home to elongated spindle neurons that, prior to 2007, were known only in hominids. In humans, these cells are involved in higher-order cognitive and socioemotional abilities such as social judgment, emotional understanding, and theory of mind. Cetacean spindle neurons are found in areas of the brain that are homologous to where they are found in humans suggesting that they perform a corresponding function.

Even so, language comprehension and production in humans is not like a computer translating or mapping sounds onto dictionary meanings that are then regurgitated on request. For one thing, language and gesture often accompany each other in parallel, whether in dogs <bark + action> or people <verbal action + gesture>. David McNeill, a psycholinguist at the University of Chicago, has demonstrated that speech and gesture, although independent channels, are closely associated. Gesture, however, is an imagistic and global-synthetic communicative channel whereas language is a syntactic and linear-segmented one. The field that studies the use of signs or symbols to convey meaning, such as words in speaking, gestures as in a goodbye wave, and a musical leitmotif that signifies a particular character in an opera, is known as *semiotics*. Gesture and speech are different semiotic channels for communicating information to conspecifics or others. According to Professor McNeill, the global-synthetic image appears first—that is, gesture anticipates speech—and then metamorphizes into a complex imagistic-syntactic structure. Interestingly, metaphoric gesture, a gesture that stands in place of something else as in two fingers signifying victory, functions to create a relationship between an image (V sign) and an abstract concept (victory). Drawing is a good example. Drawings essentially function as metaphoric gestures on paper, according to Rudolf Arnheim, the former psychologist of art. That is to say, drawings emanate from gestures of the body with the combined operations of the hand, shoulder, and torso.

> The perceptual qualities of shape and motion are present in the very acts of thinking depicted by the gestures.[13]

On the other hand, animal-produced sound is usually embedded within a social context and in most sound-producing animals largely has a social function. For example, there is evidence that vocal mimicry in birds or copying other bird's sounds, evolved from song learning or copying other bird's songs, and that nonvocal mimicry (preening one's feathers with other birds), which has a narrower purview in communication, actually evolved from vocal mimicry. Bird sound, song, and mimicry are thus socially acquired and have adaptive functions such as finding mates and defending territoriality. Imitation of movement in mammals is a probably a separate evolutionary development. We might thus step back a moment and distinguish among emulation, mimicry, and imitation.

> … imitation is reserved for instances when an animal performs an otherwise improbable action as a result of seeing another animal perform it. Goal emulation (or just "emulation") occurs when the animal learns from observation where to direct its behavior or what goal is to be achieved but does not copy precisely the actions of the demonstrator.[14]

Recent research has shown that Egyptian fruit bat pups, *Rousettus aegyptiacus*, engage in vocal mimicry and that such mimicry is probably more widespread in the animal kingdom than previously thought. This particular subspecies of bat is an unusually vocal and social mammal. They live in large colonies of up to thousands of individuals and they hear sequences of multiharmonic calls throughout their lives. Experimental manipulation of mothers' vocalizations by altering the fundamental frequency of her utterances (F0, the lowest frequency of a periodic waveform as in a musical pitch such as middle C) resulted in the pups producing distinct dialects. The fundamental frequency is high in pups (8–15 kHz) but decreases with age in adults (0.2–1.2 kHz). In the case of fruit bats, but not songbirds, bats learn vocal sounds not just from their mothers but from *crowd-sourced vocalizations* suggesting a distinct resemblance to vocal learning in other species including cetaceans, elephants, marmoset monkeys, pinnipeds (sea lions, seals, walruses), and of course, modern hominins.

Similarly, barking in canines has important social functions including establishing territoriality and signaling a need for food, mates, play, and the like. As you may remember, canines have lived in close proximity to humans for 30,000–40,000 years or more based on genetic and evolutionary evidence and appear to have originally evolved from Old World gray wolves. New World dogs likely accompanied late Pleistocene hominins across the Bering Strait into North America. The enculturation of dogs and humans together has led to communicative abilities in canines that are superior in some ways to infra-human primates. Canines, for instance, are better at locating hidden food than chimpanzees when a human provides a communicative, nonverbal cue such as pointing to, looking at, tapping on its location, or placing a mark on the hidden food's container. Even canine puppies as young as nine weeks are better than chance at performing the very same task. Culture, writ large, has always played a hugely significant role in the evolution of the MBB.

Besides its obvious social utility in canines, nonverbal or bodily-based communication between dogs is a vitally important aspect of their canine ecosystem. Outside of communicative sounds such as barking, much of canine communication involves eye gaze, facial expression (bared teeth), posture, as well as gesture including the position of the head, legs, paws, and tail. Canines may signal aggression by growling and baring their teeth or assume a submissive posture when requesting affection or play. Canines also make note of changes in position of movements with regard to orientation of the body and head. They even make use of knitted eyebrow displays when trying to figure out a problem and in a similar manner to humans. This is accomplished by the same *corrugator supercilii muscle* found in us that pulls the eyebrows both downward and outward and creates the classic vertical wrinkles of the forehead in humans. Tail-wagging, too, plays a very important role in communication in dogs and is fully established in most breeds by 49 days of age or seven weeks. It signals sociality or interest, among other things, as well as communicates object location in some breeds by physically pointing with the head, tail, and body. All that being said, the supreme importance of the body in communicating intention should never be underestimated.

The Edge of Evolution

The first adaptation that was directly shaped by evolution was *dichotomization* for representing in a new cell structure, the neuron, physically discrete and opposed features of the world such as light/dark, hot/cold, smooth/rough, near/far, and big/small. In order to encode features of the external world into internal properties of a biological system, very early Animalia began to "parse natural variety [in the external world] into pairs of opposed and dichotomous categories,"[15] which then became an inherent property of animal mentation throughout evolution. From this inception point, minds and brains evolved. Continuous variables in the natural world are potentially infinite in the amount of information they can hold, so discretization—transforming continuous physical variables or features into individuated ones—simplified the process of perceiving and categorizing features of nature by a biological system. Not surprisingly, an analogous binary code is instantiated in the modern digital computer. Of course, in machines and living beings there may be alternatives ways to parse the natural or the physical world. But this is what nature on our planet ultimately evolved.

Most of what we normally assume that the brain does well were not actually selected for by evolution. Let's see: Reading, writing, mathematical calculation, interpreting a piece of art, remembering a phone number, the concept of love, taking a jump shot, tennis, driving a car, skipping, making dinner in a microwave, and so on. Areas of the brain and body like the infra-parietal lobule for mathematical calculation and the motor areas for using a hammer with your hand, were not primary adaptations shaped by evolution for these specific uses but were useful features of the MBB that were later coopted for some new cognitive trait like counting or building a hut. They were *exaptations*, not adaptations, and consequently they were not directly shaped by evolution. An upright, bipedal posture was a primary adaptation but use of the hands for creating something with the aid of tools (manual dexterity), was not. To put it another way, an upright posture functions as an invariant, historical constraint on subsequent exaptations that arose from an ancestral state that represented fundamental or *plesiomorphic features* of a common ancestor and its descendants. Stephen Jay Gould, the former Harvard paleontologist, contends that the "brain harbors more cooptable spandrels than primary adaptations."[16] Much of the cognitive abilities of modern humans are not primary adaptations, but exaptations for current use like reading, writing, and arithmetic. It's variety and flexibility that really matter.

What was the common ancestor of modern humans (Figure 3.3)? It appears that African apes or *hominids* (orangutans, gorillas, chimpanzees, and archaic humans) began to diverge from each other somewhere around 5–7 million years ago or earlier. It apparently was not a clear split, but a messy breakup over an extended evolutionary period, as these evolving species may have interbred for a considerable period of time over approximately one million years or so. Some of the brightest, non-human primates are the bonobos, which are members of the chimpanzee family. They communicate vocally with their kin but use similar calls or vocal signs to mean diverse things so that the receiving bonobo must determine the meaning from the context, a characteristic of modern life not at all unfamiliar to us. They possess the capacity for self-awareness as they easily pass the mirror self-recognition test. That is, they can distinguish between themselves and others while looking into a mirror. But they also display keen interest in aspects of themselves such as a red mark placed surreptitiously on their face and the assumption is that this behavior exemplifies self-awareness. So, it turns out we may not be the only creatures on the planet that possess self-awareness or higher-order or secondary consciousness. Their facial expressions and hand gestures are very similar to humans and, what's more, are easily understood by us suggesting that both species—bonobos and humans—are picking up on intentional states conveyed by way of the body. They engage in observational learning by watching others, including other humans, a sophisticated way to learn that young children and a few other species quickly pick up on. But, most extraordinarily, they have been taught up to 3,000 words using geometric-like symbols called lexigrams. They can communicate with about 500 or so of them, as they cannot intone human word sounds themselves, but can respond with these lexicographical tokens to spoken sentences in communication with others, including humans.

One common ancestor that has been suggested that predated African apes was a hominid species known as *Orrorin tugenensis* or "Original Man" whose partial remains were found in the year 2000 in Kenya and date back 5.7 to 6.1 million years ago. It may have been bipedal engaging in an upright walking posture, but it was also prehensile in that it probably climbed as its skeletal remains suggested it had appendages for grasping at things—even toolmaking—which would suggest advanced cognitive abilities. It had small teeth closer to those of modern humans than other hominoid species and it lived in an arid evergreen forest habitat based on the fossils of other flora and fauna found beside it.

Sahelanthropus tchadensis was an even earlier hominid species that lived around 7 million years ago but predated Orrorin. Little is known about this species, however, because only a skull was recovered without any skeletal remains. The cranial remains are much different from modern humans. The canine teeth are unusually small. Canines are teeth on each side of the incisors of which the latter make up the four front teeth in both the upper and lower jaw in modern humans. The layout of the teeth in *Sahelanthropus tchadensis* is U-shaped but more like Australopithecines, an earlier hominid. There are heavy brow ridges and the face is flatter than modern humans. Yet, there is an anterior foramen magnum, where the spinal cord passes through into the cranial cavity, suggesting that they may have been bipedal walkers. When the foramen magnum is shifted forward as it is in *S. tchadensis* it suggests that they were bipedal as is true in modern humans as well as other early bipedal hominids such as *Australopithecus africanus*.

Australopithecines, of which *Australopithecus africanus* is one species, were quite advanced in their behavioral and cognitive abilities long before modern humans appeared on Planet Earth. They arrived on the evolutionary scene about 3.5 million years ago or earlier. Paleontological evidence suggests that they were bipedal walkers, made use of stone tools—particularly for carving up animal carcasses for meat—and may have even made containers for food and water and thus had some imitative abilities to construct these assorted objects. It also suggests that they may have possessed vestigial elements of symbolic thought, incipient self-awareness of themselves and others, and even some early form of human speech. Nonetheless, they had relatively graceful bodies on an ape-like frame, which is one reason they were named *Australopithecus* or ape-like hominins.

A second punctuated burst in evolution around 400,000 years ago (400ka) suggests a significant increase in brain size occurred in hominins.

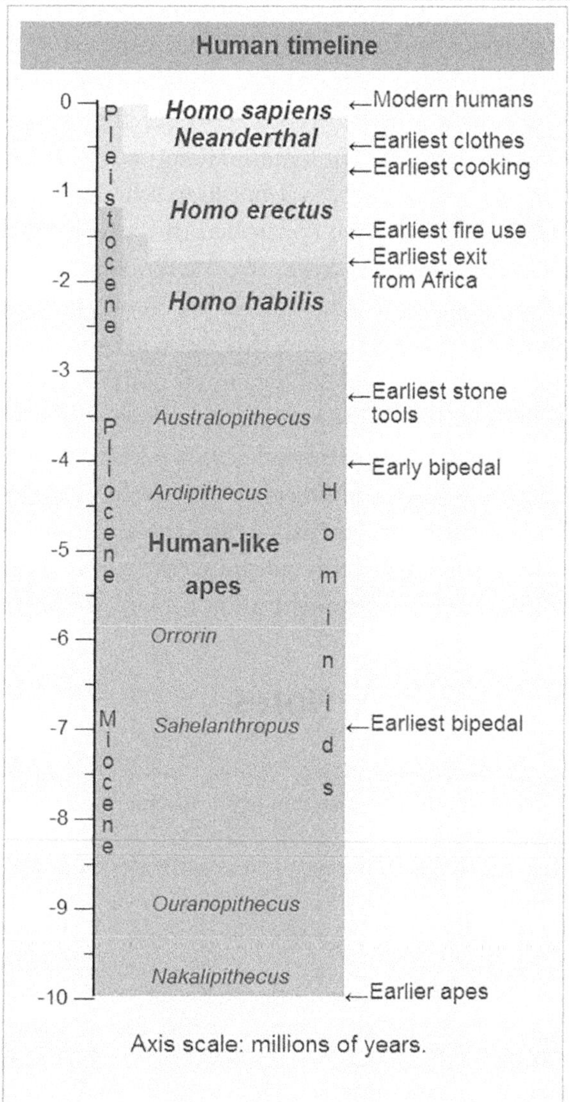

Figure 3.3. Human Timeline.
Credit: https://en.wikipedia.org/wiki/Template:Human_timeline, "Human Timeline," https://creativecommons.org/licenses/by-sa/3.0/legalcode. The original work has been modified.

With the advent of fire, around a million-and-a-half years ago, cooked food, diet, and nutrition rapidly improved bringing on the benefits to fostering increasing brain size. The *dexterity hypothesis* discussed in subsequent chapters would have brought not only the increase in *manual dexterity* millions of years earlier—even before the first punctuated burst in early hominins

1.7 million years ago that increased body size—but later *oral dexterity* in consuming cooked foods and *vocal dexterity* that ushered in speech and language. All of this was enabled by the neocortex of early humans getting direct throughput to the neurons of the motor cortices that controlled the jaws, tongue, face, and vocal cords for oral, manual, and vocal dexterity. As the lateral prefrontal cortex (LPFC) evolved it began to send its neural projections to the spinal cord and medulla, which controlled the speech articulatory apparatus (larynx, tongue, the upper and lower lip, the hard and soft palate, pharynx, and epiglottis). Without this motor articulatory network, speech would have never evolved in modern humans. But this network also controlled communication through use of the hands and manual gestures as the two—gesture and speech—are closely intertwined within the brain and body and with each other in evolution. The lateral prefrontal cortex is central to decision-making and taking novel approaches to solving problems in the environment that early humans undoubtedly came up against. Thus, these various dexterities were enlisted as a whole to solve problems central to the survival of early humans.

We'll take up the nature of thought and problem-solving next.

Notes

1. The quote, "Self-generated movement …" is from Frank R. Wilson, *The Hand*, 1998.
2. The quote, "Psychology will be based on a new foundation …" is from Charles Darwin, *The Origin of Species*, 1859.
3. The quote, "Nature is prodigal in variety …" is from Charles Darwin, *The Origin of Species*, 1859.
4. The quote, "Praxis … is … the heterogeneous class …" is from Frank R. Wilson, *The Hand*, 1998.
5. The quote, "… is a form of reasoning …" is from Rudolf Arnheim, *Visual Thinking*, 1969.
6. The quote, "… 'visual thinking' means that …" is from Rudolf Arnheim, *Visual Thinking*, 1969.
7. The quote, "Language, on the other hand …" is from Grundmann & Arnheim, *The Intelligence of Vision: An Interview with Rudolf Arnheim*, 1998.
8. The quote, "Imagine that a mad scientist takes out your brain …" is from Daniel Toker, *We are All Brains in Vats*, 2017, June 28.
9. The quote, "Everything can be taken from a man …" is from Viktor E. Frankl, *Man's Search for Meaning*, 1946.
10. The quote, "No mere summing up of the separate actions …" is from J. S. Mill, *A System of Logic*, 1843.
11. The quote, "The cerebellum is intimately linked …" is from D. R. Hampson & G. J. Blatt, *Autism spectrum disorders and neuropathology of the cerebellum*, 2015.

12. The quote, "Even in the simple form of Pavlovian conditioning ..." is from S. J. Shettleworth, *Cognition, Evolution, and Behavior*, 1998.
13. The quote, "The perceptual qualities of shape ..." is from Rudolf Arnheim, *Visual Thinking*, 1969.
14. The quote, "... imitation is reserved for instances ..." is from S. J. Shettleworth, *Cognition, Evolution, and Behavior*, 1998.
15. The quote, "parse natural variety [in the external world] ... is from S. J. Gould, *The Structure of Evolutionary Theory*, 2002.
16. The quote, "brain harbors more cooptable spandrels ..." is from S. J. Gould, *The Structure of Evolutionary Theory*, 2002.

Bibliography

Akshoomoff, N. A., & Courchesne, E. (1994). ERP evidence for a shifting attention deficit in patients with damage to the cerebellum. *Journal of Cognitive Neuroscience*, 6(4), 388–399.

Andics, A., Gabor, A., Gasci, M., Farago, T., Szabo, D., & Miklosi, A. (2016). Neural mechanisms for lexical processing in dogs. *Science*, *353*(6303), 1030–1032.

Andreasen, N. C., Paradiso, S., & O'Leary, D. S. (1998). "Cognitive dysmetria" as an integrative theory of schizophrenia: A dysfunction in cortical-subcortical-cerebellar circuitry? *Schizophrenia Bulletin*, *24*(2), 203–218.

Angier, N. (2017, September 22). Birds beware: The praying mantis wants your brain. *The New York Times*, pp. 1–7. Retrieved from https://www.nytimes.com/2017/09/22/science/praying-mantis-eating-birds.html

Arden, R., Bensky, M. K., & Adams, M. J. (2016). A review of cognitive abilities in dogs, 1911 through 2016: More individual differences, please! *Current Directions in Psychological Science*, *25*(5), 307–312.

Argyle, M. (1989). *Bodily communication*. (2nd ed.). London: Methuen.

Arnheim, R. (1969). *Visual thinking*. Berkeley, CA: University of California Press.

Arnheim, R. (1998, August-September). Rudolf Arnheim: "Die Intelligenz des Sehens," *Neue Bildende Kunst*, pp. 56–62.

Baldwin, J. M. (1896). A new factor in evolution. *The American Naturalist*, *30*(354), 441–451.

Barten, S. S. (1979). Development of gesture. In N. R. Smith & M. B. Franklin (Eds.), *Symbolic functioning in childhood* (pp. 139–151). Hillsdale, NJ: Lawrence Erlbaum.

Bartlett, F. C. (1958). *Thinking: An experimental and social study*. New York, NY: Basic Books.

Bates, E., Thal, D., Whitesell, K., Fenson, L., & Oakes, L. (1989). Integrating language and gesture in infancy. *Developmental Psychology*, *25*(6), 1004–1019.

Bekoff, M. (1998). Playing with play: What can we learn about cognition, negotiation, and evolution? In D. D. Cummins & C. Allen (Eds.), *The evolution of mind* (pp. 162–182). Oxford, NY: Oxford University Press.

Bekoff, M. (2002). *Minding animals: Awareness, emotions, and heart*. Oxford, NY: Oxford University Press.

Bracke-Tolkmitt, R., Linden, A., Canavan, A. G. M., Rockstroh, B., Scholz, E., Wessel, K., & Diener, H.-C. (1989). The cerebellum contributes to mental skills. *Behavioral Neuroscience, 103*(2), 442–446.

Bruner, J. (1983). *Child's talk: Learning to use language*. New York, NY: W. W. Norton.

Buzsaki, G. (2006). *Rhythms of the brain*. Oxford, NY: Oxford University Press.

Buzsaki, G., & Draguhn, A. (2004). Neuronal oscillations in cortical networks. *Science*, 304, 1926–1929.

Chomsky, N. (2011). Language and other cognitive systems: What is special about language? *Language Learning and Development*, *7*(4), 263–278.

Corballis, M. (1999). The gestural origins of language. *American Scientist*, *87*(2), 138–145.

Corballis, M. (2002). *From hand to mouth: The origins of language*. Princeton, NJ: Princeton University Press.

Corballis, M. (2011). *The recursive mind: The origins of human language, thought, and civilization*. Princeton, NJ: Princeton University Press.

Corballis, M. (2017). *The truth about language: What it is and where it came from*. Chicago, IL: The University of Chicago Press.

Coren, S. (2000). *How to speak dog: Mastering the art of dog-human communication*. New York, NY: Free Press.

Courchesne, E., Townsend, J., Akshoomoff, N. A., Saitoh, O., Yeung-Courchesne, R., Lincoln, A. J., … Lau, L. (1994). Impairment in shifting attention in autistic and cerebellar patients. *Behavioral Neuroscience*, *108*(5), 848–865.

Crammond, D. J. (1997). Motor imagery: Never in your wildest dream. *Trends in Neurosciences*, *20*(2), 54–57.

Damasio, A. R. (1989). Time-locked multiregional retroactivation: A systems level proposal for the neural substrates of recall and recognition. *Cognition*, *33*(1–2), 25–62.

Damasio, A. R. (1996). The somatic marker hypothesis and the possible functions of the prefrontal cortices. *Philosophical Transactions of the Royal Society of London (Series B)*, *351*(1346): 1413–1420.

Damasio, A. R. (1999). *The feeling of what happens: Body and emotion in the making of consciousness*. New York, NY: Harcourt.

Damasio, A. R. (2003). *Looking for Spinoza: Joy, sorrow, and the feeling brain*. New York, NY: Harcourt.

Damasio, A. R., & Damasio, H. (1994). Cortical systems for retrieval of concrete knowledge: The convergence zone framework. In C. Koch (Ed.), *Large-scale neuronal theories of the brain* (pp. 61–74). Cambridge, MA: MIT Press.

Darwin, C. (1859). *The origin of species*. London: Penguin Books.

Darwin, C. (1879). *The descent of man and selection in relation to sex*. (2nd ed.). London: Penguin Books.

Darwin, C. (1965). *The expression of the emotions in man and animals*. Chicago, IL: The University of Chicago Press. (Original work published 1872).

Dehaene, S. (2009). *Reading in the brain: The new science of how we read*. London: Penguin Books.

Dehaene, S. (2014). *Consciousness and the brain: Deciphering how the brain codes our thoughts*. New York, NY: Penguin Books.

DeLong, M. R. (1990). Primate models of movement disorders of basal ganglia origin. *Trends in the Neurosciences, 13*(7), 281–285.

Dewey, D., & Tupper, D. E. (2004). (Eds.). *Developmental motor disorders: A neuropsychological perspective*. New York, NY: The Guilford Press.

Edelman, G. M. (1990). *The remembered present: A biological theory of consciousness*. New York, NY: Basic Books.

Ekman, P., & Friesen, W. V. (1969). The repertoire of nonverbal behavior: Categories, origins, usage, and coding. *Semiotica, 1*(1), 49–98.

Eldredge, N., & Gould, S. J. (1972). Punctuated equilibria: An alternative to phyletic gradualism. In T. J. M. Schopf (Ed.), *Models in paleobiology* (pp. 82–115). San Francisco, CA: Freeman, Cooper.

Erdal, D., & Whiten, A. (1996). Egalitarianism and Machiavellian intelligence in human evolution. In P. Mellars & K. Gibson (Eds.), *Modelling the early human mind* (pp. 139–150). Cambridge, UK: McDonald Institute for Archaeological Research.

Farahany, N. A., Greely, H. T., Hyman, S., Koch, C., Grady, C., Pasca, S. P., … Song, H. (2018, April 26). The ethics of experimenting with human brain tissue. *Nature*, 556(7702), 429–434.

Feinberg, T. E., & Mallatt, J. M. (2016). *The ancient origins of consciousness: How the brain created experience*. Cambridge, MA: MIT Press.

Fiez, J. A., Petersen, S. E., Cheney, M. A., & Raichle, M. E. (1992). Impaired non-motor learning and error detection associated with cerebellar damage: A single case study. *Brain, 115*, 155–178.

Frankl, V. E. (1946). *Man's search for meaning*. New York, NY: Beacon Press.

Fugazza, C., Pogany, A., & Miklosi, A. (2016, November 23). Recall of other's actions after incidental encoding reveals episodic-like memory in dogs. *Current Biology*, 26(23), 3209–3213.

Goffman, E. (1961). *Asylums: Essays on the social situation of mental patients and other inmates*. New York, NY: Anchor Books.

Goldin-Meadow, S. (1997). When gestures and words speak differently. *Current Directions in Psychological Science*, 6(5), 138–143.

Gould, S. J. (1977a). *Ontogeny and phylogeny*. Cambridge, MA: Harvard University Press.

Gould, S. J. (1977b). *Ever since Darwin: Reflections in natural history*. New York, NY: W. W. Norton.

Gould, S. J. (1981). *The mismeasure of man*. New York, NY: W. W. Norton.

Gould, S. J. (2000, June 25). The brain of brawn. *The New York Times*, p. A20. Retrieved from https://www.nytimes.com/2000/06/25/opinion/the-brain-of-brawn.html

Gould, S. J. (2002). *The structure of evolutionary theory*. Cambridge MA: Belknap Press of Harvard University Press.

Gould, S. J., & Lewontin, R. (1979). The spandrels of San Marco and the Panglossian paradigm: A critique of the adaptationist programme. *Proceedings of the Royal Society of London, B, Biological Sciences*, *205*(1161), 581–598.

Gould, S. J., & Vrba, E. (1982). Exaptation—a missing term in the science of form. *Paleobiology*, 8(1), 4–15.

Grundmann, U., & Arnheim, R. (2001). *The intelligence of vision: An interview with Rudolf Arnheim*, 2, [online]. Retrieved from http://www.cabinetmagazine.org/issues/2/rudolfarnheim.php

Gu, Z., Kalambogias, J., Yoshioka, S., Han, W., Li, Z., Kawasawa, Y., … Yoshida, Y. (2017). Control of species-dependent cortico-motoneuronal connections underlying manual dexterity. *Science*, *357*(6349), 400–404.

Hampson, D. R., & Blatt, G. J. (2015). Autism spectrum disorders and neuropathology of the cerebellum. *Frontiers in Neuroscience*, 9, 420. doi:10.3389/fnins.2015.00420

Hauser, M. D. (2000). *Wild minds: What animals really think*. New York, NY: Henry Holt.

Hazlett, H. C., Gu, H., Munsell, B. C., Kim, S. H., Styner, M., Wolff, J. J., … Zwaigenbaum, L. (2017). Early brain development in infants at high risk for autism spectrum disorder. *Nature*, *542*(7641), 348–351.

Hebb, D. O. (1949). *The organization of behaviour*. New York, NY: John Wiley & Sons.

Heilman, K. M., Rothi, L. J., & Valenstein, E. (1982). Two forms of ideomotor apraxia. *Neurology*, *32*(4), 342–346.

Hewes, G. W. (1973). Primate communication and the gestural origin of language. *Current Anthropology*, *14*(1–2), 5–24.

Hockett, C. (1960). The origin of speech. *Scientific American*, *203*(3), 88–111.

Hof, P. R., Chanis, R., & Marino, L. (2005). Cortical complexity in cetacean brains. *The Anatomical Record (Part A)*, *287A*, 1142–1152.

Horowitz, A. (2011). Theory of mind in dogs? Examining method and concept. *Learning and Behavior*, *39*(4), 314–317.

Horowitz, A. (2017). Smelling themselves: Dogs investigate their own odours longer when modified in an "olfactory mirror" test. *Behavioral Processes*, *143*, 17–24.

Hughes, H. C. (1999). *Sensory exotica: A world beyond human experience*. Cambridge, MA: MIT Press.

Humphrey, N. (1976). The social function of intellect. In P. P. G. Bateson & R. A. Hinde (Eds.), *Growing points in ethology* (pp. 303–317). Cambridge, UK: Cambridge University Press.

Hurley, S., & Chater, N. (2005). *Perspective on imitation: From neuroscience to social science*. Cambridge, MA: MIT Press.

Iacoboni, M. (2005). Neural mechanisms of imitation. *Current Opinion in Neurobiology*, *15*(6), 632–637.

Iacoboni, M., Woods, R. P., Brass, M., Bekkering, H., Mazziotta, J. C., & Rizzolatti, G. (1999). Cortical mechanisms of human imitation. *Science*, *286*(5449), 2526–2528.

Imamizu, H. (2010). Prediction of sensorimotor feedback from efference copy of motor commands: A review of behavioral and functional neuroimaging studies. *Japanese Psychological Research*, *52*(2), 107–120.

Iverson, J. M., & Goldin-Meadow, S. (1998). Why people gesture when they speak. *Nature*, *396*(6708), 228.

Jones, S. (2005). The role of mirror neurons in imitation. In S. Hurley & N. Chater (Eds.), *Perspectives on imitation: From neuroscience to social science*. Vol. 1. Mechanisms of imitation and imitation in animals (pp. 205–210). Cambridge, MA: MIT Press.

Kaminski, J., Call, J., & Fisher, J. (2004). Word learning in the domestic dog: Evidence for "fast mapping." *Science, 304*(5677), 1682–1683.

Kaminski, R. M., Marini, H., Ortinski, P. I., Vicini, S., & Rogawski, M. A. (2006). The pheromone androstenol (5 alpha-androst-16-en-3 alpha-ol) is a neurosteroid positive modulator of GABAA receptors. *Journal of Pharmacological Experimental Therapeutics, 317*(2), 694–703.

Kennedy, J. M. (1997). How the blind draw. *Scientific American, 276*(1), 76–81.

Locke, J. (1894). *An essay concerning human understanding*. (A. C. Fraser, Ed., 2 vols.). Oxford, UK: Clarendon Press.

Marr, D. (1969). A theory of cerebellar cortex. *Journal of Physiology, 202*(2), 437–470.

Marr, D. (1982). *Vision: A computational investigation into the human representation and processing of visual information*. New York, NY: Freeman.

McNeill, D. (1985). So, you think gestures are nonverbal? *Psychological Review, 92*(3), 350–371.

McNeill, D. (1989, April). *Contribution to the "When is gesture like language?" Symposium*. Paper presented at the Biennial Meeting of the Society for Research in Child Development, Kansas City, MO.

McNeill, D. (1992). *Hand and mind: What gestures reveal about thought*. Chicago, IL: The University of Chicago Press.

Meltzoff, A. N., & Decety, J. (2003). What imitation tells us about social cognition: A rapprochement between developmental psychology and cognitive neuroscience. *Philosophical Transactions of The Royal Society, B, Biological Sciences, 358*(1431), 491–500.

Meltzoff, A. N., & Moore, M. K. (1997). Explaining facial imitation: A theoretical model. *Early Development & Parenting*, 6(3–4), 179–192.

Muto, V., Jaspar, M., Meyer, C., Kusse, C., Chellappa, S. L., Degueldre, C., … Maquet, P. (2016). Local modulation of human brain responses by circadian rhythmicity and sleep debt. *Science, 353*(6300), 687–690.

Nityananda, V., Tarawneh, G., Rosner, R., Nicolas, J., Crichton, S., & Read, J. (2016). Insect stereopsis demonstrated using a 3D insect cinema. *Science Reports*, (6), 18718.

Panksepp, J. (2005). Affective consciousness: Core emotional feelings in animals and humans. *Conscious Cognition, 14*(1), 30–80.

Piaget, J. (1952). *The origins of intelligence in children*. (2nd ed., M. Cook, Trans.). New York, NY: International Universities Press.

Prat, Y., Azoulay, L., Dor, R., & Yovel, Y. (2017). Crowd vocal learning induces vocal dialects in bats: Playback of conspecifics shapes fundamental frequency usage by pups. *PLoS Biology, 15*(10), 1–14.

Premack, D., & Premack, A. (2003). *Original intelligence: Unlocking the mystery of who we are*. New York, NY: McGraw-Hill.

Ratey, J. J. (2008). *Spark: The revolutionary new science of exercise and the brain*. New York, NY: Little, Brown.

Ridley, M. (2004). *Evolution*. (3rd ed.). Oxford, UK: Blackwell Publishing.

Roth, G., & Dicke, U. (2005). Evolution of the brain and intelligence, *Trends in the Cognitive Sciences*, 9(5), 250–257.

Rozin, P. (1976). The evolution of intelligence and access to the cognitive unconscious. In J. M. Sprague & A. N. Epstein (Eds.), *Progress in psychobiology and physiological psychology* (vol. 6, pp. 245–280). Cambridge, MA: Academic Press.

Saito, A. (1996). Social origins of cognition: Bartlett, evolutionary perspective, and embodied mind approach. *Journal of the Theory of Social Behavior, 26*(4), 399–421.

Schmahmann, J. D., & Sherman, J. C. (1998). The cerebellar cognitive affective syndrome. *Brain, 121*, 561–579.

Seitz, J. A. (1996). *Social deficits in autistic children: The role of emotion and empathy*. Unpublished manuscript, Department of Political Science & Psychology, York College, City University of New York (CUNY).

Seitz, J. A. (1999, June 4). Selfish sentinels in cooperative mammals [Letter to the editor: Darwinian selfishness. Response to T. H. Clutton-Brock]. *Science, 284*(5420), 1640–1644.

Shettleworth, S. J. (1998). *Cognition, evolution, and behavior*. Oxford, NY: Oxford University Press.

Smith, E. E., & Jonides, J. (1999). Storage and executive processes in the frontal lobes. *Science, 283*(5408), 1657–1661

Sober, E., & Wilson, D. S. (1997). *Unto others: The evolution and psychology of unselfish behavior*. Cambridge, MA: Harvard University Press.

Stiles, J. (2008). *The fundamentals of brain development: Integrating nature and nurture*. Cambridge, MA: Harvard University Press.

Striedter, G. F. (2005). *Principles of brain evolution*. Sunderland, MA: Sinauer Associates.

Sussman, A., & Hollander, J. B. (2015). *Cognitive architecture: Designing for how we respond to the built environment*. New York, NY: Routledge.

Tyler, C. W. (2018, October 18). Evidence that Leonardo da Vinci had strabismus. *JAMA Ophthalmology*, [online: doi:10.1001/jamaophthalmol.2018.3833].

Vitzthum, H., Müller, M., & Homberg, U. (2002). Neurons of the central complex of the locust *Schistocerca gregaria* are sensitive to polarized light. *The Journal of Neuroscience, 22*(3), 1114–1125.

Welsh, J. P., & Llinas, R. (1997). Some organizing principles for the control of movement based on olivocerebellar physiology. *Progress in Brain Research, 114*, 449–461.

Wilson, F. R. (1999). *The hand: How its use shapes the brain, language, and human culture*. New York, NY: Knopf, Doubleday.

Windridge, D., & Thill, S. (2018). Representational fluidity in embodied (artificial) cognition. *Biosystems, 172*, 9–17.

Wynne, C. D. L. (2001). *Animal cognition: The mental lives of animals*. New York, NY: Palgrave.

· 4 ·

THOUGHT

Physical intelligence includes conscious and nonconscious mechanisms and nonconscious mental operations of athletes are not a lesser form of intellect.[1]

Numeracy

A single structural principle underlies the relationship among the hands and body, spatial relationships deduced by the brain from the external world—that is to say, an abstract framework created in the cerebral cortex consisting of a cognitive map of the layout of objects in the environment—and an imaginary number line that extends out in two-dimensional space. The spatial layout of the fingers of the hand in the real world and topographically in the cerebral cortex and their numerical representation on a number line are all structurally linked. For one thing, there is evidence that the inferior parietal cortex (IPC) is made up of regions for the representation of the fingers, space, writing, and numbers. For another thing, numerical magnitude, physical size, and location are closely related in the brain. Indeed, when adding up numbers our attention moves to the right; when subtracting numbers our attention moves to the left.

Imagine a segment of a number line stretched out from left-to-right in which each segment is laid out continuously in space and with each location on the number line representing a certain quantity. This imaginary number line in our mind is actually the result of the coding of both spatial and numerical information within the same part of the cerebral cortex and is even influenced by the direction of writing (*Nota bene*: only a few languages are written right-to-left in their native script). Similarly, when encoding numbers verbally, reading numbers, and memorizing numbers from the printed page our mind-brain-body (MBB) uses the same coding procedures. Exposure to sex hormones in utero as well as the influence of genes can influence brain organization and the learning of numbers and space in childhood. While human infants do not differ by gender in arithmetic ability, during the preschool years, females tend to become better at calculation and males tend to become better at mathematical problem-solving.

There is a cortical-subcortical loop from the left hemisphere to the basal ganglia in the subcortex. It is involved in the reproduction of motor and verbal sequences that have become automatic such as the recitation of multiplication tables or pieces of poetry or prose, a nursery rhyme or a short story. Planning and ordering, correction of errors, and decision-making as in "What poem should I try to recite next to soothe her heart?" are carried out by the regions in the prefrontal and anterior cingulate cortices including the dorsal lateral prefrontal cortex (DLPFC), which is involved in strategy and planning. Meanwhile, the left basal ganglia and thalamus are marshaled for arithmetic memory, areas of the inferior parietal cortex for the representation of quantity, adjacent areas of the temporal and occipital cortices for the visual identification of digits and words, and the auditory cortex in the left hemisphere for the verbal comprehension and production of words and numbers. The inferior parietal cortex is also responsible for converting numbers into actual quantities as well as representing relative magnitude on our hypothetical number line, for instance, 2 and 5; 6 and 9. Exact numbers are coded in the left parietal cortex in an area called the angular gyrus (BA39 in the anterolateral region of the parietal lobe) while approximate magnitudes are represented in more expansive regions of the parietal cortex.

As the mind-brain-body (MBB) evolved, areas of the brain and body that were evolutionarily older were converted for current use towards the manipulation of mathematical symbols such as integers, operands, functions, complex numbers, relations, combinatorics, statistical and probabilistic operators, constants, and the like. This is termed, when applied to the brain, *neuronal recycling theory*.

The hIPS or horizontal part of the intraparietal sulcus (the ridges on the lateral surface of the parietal cortex), has some intriguing properties. It integrates all representations of a number quantity such as the spoken word ("four"), its Arabic number (4), its symbol, and its representation in spatial dots (: :); hence, it is multimodal or captured in multiple sensory modalities. Neurologically, when we are actually judging a numerical magnitude—Is 9 bigger than 6?—the hIPS begins to expand to neighboring brain regions to get a sense of size, distance, and location and this is all an approximation as far as the brain and body (eye and hand movements, spatial distance and location) are concerned. Or, to put it another way, the MBB uses this information to guide both initial reaching as well as coordinated hand and body movements.

A rare condition, arithmetic epilepsy or *epilepsia arithmetics*, occurs when a patient experiences an epileptic fit and produces periodic rhythmic discharges of the body while solving a mathematics problem. Electroencephalogram (EEG) recordings have pinpointed the inferior parietal cortex as the area of the brain that is convulsing. Similarly, *developmental dyscalculia* or problems in understanding numbers and their operations in childhood, is associated with loss of grey matter (the thinking part of the brain) in the intraparietal sulcus.

Mathematics

Our mathematical conceptual system is, not surprisingly, also grounded in basic sensorimotor experiences. For instance, arithmetic consists of object collection (numbers as collection of physical objects), object construction (quantity), and physical motion (number locations as situated on a path or continuum). The parietal lobes and the intraparietal sulci shape the neural circuit that underlies handshapes and finger movements, which also contribute to finger counting and finger calculation. The latter two are part of a universal cross-cultural stage in the learning of numbers by children all over the world and presumably since the beginning of human history as I described earlier. It indicates that the cortical representation of finger movements and numbers occupies an interrelated neural circuit that includes the spatial layout of the fingers on the hand, the cerebral representation of the fingers, and the location of number on a number line and obeys similar principles of cerebral organization. This suggests a close relationship among our body maps, our spatial maps, and the representation of number on a number line.

Recent evidence indicates that *mathematical intuition* (sic; see Chapter 11) emerges from two distinct neural systems that underlie both exact and approximate arithmetic. Whereas the former is acquired in a language-like format and uses neural networks involved in word association, the latter is language independent and uses bilateral areas in the vicinity of the intraparietal sulci (Brodmann's area or BA39). These areas are active during visual guidance of eye and hand movements, mental rotation, and orientation to the environment. They are posited to be related to preverbal numerical abilities that are found in both human infants as well as other animal species.

Numerical abilities are widely distributed in the animal kingdom. Let's see: Crows, ravens (corvids); humans; spiders and scorpions, lobsters and crabs, insects, millipedes and centipedes (all arthropods, segmented invertebrates with jointed limbs); guppies; anurans such as frogs; honeybees; African lions; chimps, rhesus macaques, and monkeys; hyenas and wolves; and birds and their chicks. Anthropods, for instance, are able to perceive simple numerosity, possess basic number percepts (distinguishing, for example, between one and three objects), and engage in rudimentary numerical computation. Nonhuman animals even follow *Weber's Law* when comparing simple numerosities. That is, when they discriminate the comparative size of two or more numbers (that is, numerosity) they comprehend that the ratio of the two quantities obeys the *size effect* (4 and 6 ~ 40 and 60, but not 50 and 52), the distance effect (the distance from 10 and 5 is larger and easier to discriminate than 6 and 5), and they also possess an approximate number system just like modern humans. Male frogs will use both the number of calls (croaking sounds; no surprise there) and type of calls, to attract female conspecifics. Honeybees assess distance and count landmarks between a food source and the hive. Social mammals gauge the number of calls from conspecifics and respond based on their numerical advantage. Corvids and rhesus monkeys are able to recognize exact numerosities (3 seeds or 6 bananas, respectively) in match-to-sample tasks. The avian pallium (layers of gray and white matter that cover the upper surface of the cerebrum), which is similar to the cortex in modern humans and putatively analogous to the prefrontal cortex (executive functions, higher cognitive abilities), includes the hippocampus, hyperpallium, cortex piriformis, and the nidopallium. Within the *caudolateral nidopallium* are neurons that are activated by numerosity. In general, these neurons encode both cognitive and sensory attributes in working memory and translate this information into efferent output for motor behavior. The caudolateral nidopallium acting as the central executive of the avian brain is clearly a different structure than the mammalian neocortex but performs a strikingly similar set of functions and arose most likely by way of convergent evolution. These two similar structures in birds and mammals and are known as a *homoplasy* or a phenotypic trait shared by two or more species arising independently in evolution and not sharing a common ancestor. However, other aspects of mathematics such as geometry are encoded in the hippocampal formation in vertebrates including fish and mammals. Monkeys and human infants can distinguish numerosity across sensory modalities, such as numerical comparisons between touch and vision, suggesting an amodal understanding of number, at least in primates.

Then there is the Gerstmann syndrome. It involves deficits in the left inferior parietal region characterized by difficulties in writing, representing the fingers of the hand, distinguishing left from right, and *acalculia* or difficulties in performing numerical operations. This further strengthens the notion that the inferior parietal area is a central brain region for numerical abilities including the representation of continuous quantities, abstract maps of spatial layouts of objects in the environment, and further subdivided into microregions specialized for finger movements and graphic abilities such as writing. Studies purporting to enhance spatial-temporal reasoning skills by exposure to music (Mozart Sonata in D Major, k. 448), keyboard training, and math video games, may be tapping into this composite neural circuit. Such findings fit in well from what is known about the modular structure of the brain. A module is a set of interconnected neurons or nodes densely coupled with each other but loosely affixed to other neurons in more far flung parts of the network. Modules are able to contain pertubations (deviations of the system from a balanced state) in the network by their ability to buffer systems that preserve the stability of the overall system. So, with increasing size of the network, along with the number of modules, protects the module from pertubations more distant in the network. Indeed, the brain's circuits are highly compartmentalized receiving synaptic connections from less than 3% of neurons of a surrounding square millimeter of cortex. This indicates that neural circuits operate more efficiently by sharing information over a smaller number of units. The extent of this neural connectedness is uniform across treeshews (*Tupaia glis*, a small mammal bigger than a mouse but smaller than a rat) who possess a higher brain to body mass ratio than any other mammal including humans, prosimians, marmosets, and primates who all differ in brain size by four orders of magnitude. In the cortex, outward folds separate strongly interconnected areas while inward folds separate weakly interconnected areas. Such compact wiring ensures that connections occur with high probability between juxtaposed regions and with less probability among more distance regions and thus contain pertubations. Moreover, these neural arrangements develop early as local physical forces and mechanical properties of surrounding tissue shape neural growth and migration.

Overall, neurobiological regions of interest in numerosity and arithmetic calculation in modern humans are bilaterally represented in the parietal lobes and particularly the intraparietal sulcus (IPS), small regions of both the left and right hemispheres for specific arithmetic operations, the left frontal lobe for orchestrating mathematical calculations, and areas of the temporal lobe for numerical memory that are separate from regions for linguistic memory. Problems in numerosity and arithmetic operations, such as dyscalculia in children, may have a genetic component and arise largely in the parietal lobes.

Science: The Structure of the Natural World

Conceptualization in science may be thought of as an embodied mental activity. In the following I explicate the embodied role of categorization, imagery, and gesture.

Categorization

Four-year-olds make significant use of motion cues to categorize objects regardless of whether the objects are geometric or animal figures. Seven-year-olds and adults tend to make use of motion cues more often to categorize animal figures but not geometric figures. Similarly, infant vervets—small grey, African monkeys—use a motion-oriented categorization scheme to classify objects initially according to their actions or behavior. Studies such as this indicate that motion plays a pivotal role in concept acquisition and guides human and infrahuman primates in the both the categorization of objects and the learning of concepts across the lifespan.

Imagery

Mental practice, or the cognitive rehearsal of a physical skill in the absence of movement, includes both external mental imagery—viewing oneself from the perspective of an observer—and internal kinesthetic imagery. Meta-analytic studies indicate that the effect of mental practice over no mental practice is highly significant (effect size = 0.68, standard deviation = 0.11). Meta-analytic studies cull results across a group of similar studies to evaluate their overall effectiveness. So, in this case, the effect size of all the studies combined is fairly large and the variation around the combined means is relatively small meaning that there wasn't much variance across the studies. Most subjects performed in approximately the same range. Moreover, kinesthetic imagery is even more effective than external imagery. This is probably due to increases in muscle or kinesthetic memory in physical demanding tasks as well as spatial, temporal, and sequential aspects of the symbolic components of the image in cognition and memory. On the other hand, external imagery is more effective for long-distance runners who tend to want to dissociate the physical pain of the body as one might expect over extended distances.

Recent studies of physical imagery suggest that it is a direct analog of physical action and not in any material sense a correlate of visual perception. Traditional models of physical imagery have explained motion without regard to the physical forces that control it. Dynamic models, however, describe the forces acting on a system in motion. Johannes Kepler's theory of planetary motion is an example. Johannes Kepler formulated laws that specified that a planetary orbit is not a circle but an eclipse and that the sun is not at the center of this elliptical orbit. These laws arose in the early 1600s based on his understanding at that time and later—courtesy of Sir Isaac Newton, a mathematician and astronomer, and Gottfried Leibniz, a German philosopher—of angular momentum and universal gravitation. These universal forces of nature operated on objects in motion influencing their resistance to the aforementioned forces as well as their rate of change. Such *enactive* or *timing-responsive representations* are stored in the MBB and allow individuals to anticipate, plan, and respond directly to the dynamic world through their physical actions. Both haptic and imagistic information are incorporated into these enactive representations so that physical imagery is predictably tied to what we actually perceive in the external world through our senses.

Gesture

Studies of the acquisition of gesture in young children indicate that *iconics* or gestures that represent details of visual images, such as an encircling gesture to represent a round room, exhibit a close relationship with spoken language development in children. So, do *beats* or gestures that illustrate temporal structure of an utterance such as a rapid beat of the hand to indicate running to get on the bus. Iconics function by illustrating predicate structure, primarily verbs + adverbs and adjectives, in order to express complex aspects of ideas such as roundness. Beats function as part of the child's ability to use varying stress patterns with multimorphemic utterances, or words that consist of a collection of meanings as in "re-appear."

Deictics (pointing gestures), *emblems* (gestures which have a direct verbal translation as in a hand wave or the compound word "bye-bye"), and *gives* (typically, gestures with hands outstretched) arise prior to spoken language and do not display direct links to speech development. Thus, gestural imitation of adults by infants from 13–16 months is unaffected by accompanying language at the earliest stages of lexical development in producing and understanding words. Indeed, infants will ignore a linguistic cue if it conflicts with a modeled gesture in later stages of infancy, and in young children with relatively large vocabulary (~200 words), such speech-gesture conflict negatively affects motor performance.

Deictic gestures comprise 80% of gestures in French-English bilingual children under four-years-of-age. In adults, reports suggest that iconics and beats make up over 60% of gestures. *Iconic gestures* in adults highlight linguistic meaning through pairing movement isomorphic in form and manner of execution (hand rises upward) with language ("He crawls up the pipe"). *Metaphoric gestures* on the other hand convey nonliteral meaning by representing abstract concepts ("confinement") through the use of gesture that depicts the underlying meaning or vehicle of the metaphor (cupping the hands as to form a container). As I mentioned earlier, graphic drawings are essentially metaphoric gestures on paper, according to Rudolf Arnheim. The perceptual qualities of shape and motion in drawings are also present in the very acts of thinking depicted by the gestures and are the medium in which the thinking itself takes place. These perceptual qualities are not necessarily just visual, however. In gestures, the kinesthetic experiences of pushing, pulling, advancing, and obstructing are likely to play an important role, too.

Then we have *protogestures*. Protogestures are the early rudiments of gesture with the hands such as an infant holding their hand up to signify something like their delight at more food. They link vocal and manual activities that evolve in the postnatal period with iconic and deictic gestures that develop over the next two years of life, followed by beats in the third year and true metaphoric gestures in the fifth to sixth year. Nonetheless, other schemes of gestural development suggest that the earliest gestures are *depictive* (a one-year-old child slowly opening and closing her mouth to represent analogous movement of a matchbox), *deictic* (pointing, showing, and giving), and *enactive and recognitory gestures* (pretending to comb your hair). These are closely followed in age by *expressive* (knitted brow) and *instrumental* (extending your arms to be picked up) *gestures*. Both scenarios suggest that two kinds of thought processes are working in parallel, as I mentioned earlier: An imagistic, global-synthetic representation and a syntactic, linear-segmented one. That is to say, a visual-kinesthetic image is forged in parallel with an inner speech symbol (talking to oneself either out loud or silently) such that gesture and speech share a *computational stage* in which the separate elements combine to form a more complex cognitive structure. This *dialectical synthesis* of the two kinds of thinking suggests that thought itself occurs within a field of oppositions—or what I referred to earlier as *dichotomization*—and how it carves up the imagistic and syntactic parts is the very embodiment of everyday thinking.

The foregoing demonstrates the central importance of the role of the body in forming complex concepts and ideas. Indeed, gestures that accompany language may facilitate thought itself. For instance, people speaking on the telephone routinely gesture even though it plays no obvious role in communication. Similarly, blind speakers when speaking to a blind listener will gesture even though such behavior does not depend on either an observer or on an adult model. In studies of gesture-speech mismatch, the expression of concepts motorically before their linguistic realization may help facilitate the working out or unpacking of ideas. Gesture-speech mismatch may also have communicative value because gesture highlights aspects of the mind of the speaker that are inconsistent with his or her spoken language. In modern choreographic practice much of the communicative and conceptual power of dance is facilitated by gesture as the doyen of modern dance choreographers, Martha Graham, once remarked.

> The body is shaped, disciplined, honored, and in time, trusted. The movement becomes clean, precise, eloquent, truthful. Movement never lies. It is a barometer telling the state of the soul's weather to all who can read it. This might be called the law of the dancer's life—the law which governs the outer aspects.[2]

Studies of the drawings of individuals with congenital and acquired blindness indicate that blind individuals are able to describe motion by making use of metaphorical ways of indicating movement such as curved spokes to denote motion of a wheel. Like sighted individuals who treat brightness borders as indicators of the edges of surfaces, blind individuals treat pressure borders as indicators of surface edges. Thus, it has been suggested that both visual and haptic information are coordinated by an *amodal system*. An amodal system associates information across the sensory modalities. This system would make use of haptic information in the blind to use touch to think with just as the sighted use visual information to plan and organize visual reality.

Implications

As electronic media, video games, and telecommunication devices become increasingly popular, there is concern that such *mechanization of culture* will severely limit the use of the body. Childhood play is essential to development and traditional mediums of play—constructive, sociodramatic, and symbolic play—facilitate bodily coordination and the physical basis of concept acquisition. Physical practice in sports, chess, and music contribute to world-class performance. Depression, too, is often associated with a negatively distorted body image and restless sleep. On the other hand, is it that our brains are largely formed independent of type of sensory input? What will become of human minds in a digitized culture?

While it is generally true that physically handicapped children eventually develop normally in spite of motoric impairments, their success is largely due to their ability to compensate for their motor limitations, not because of them. The physically impaired seek out intellectual, social, and environmental stimulation and make optimal use of what motor abilities are still intact—such as visual saccades or auditory scanning of the environment—as the nervous system reroutes degenerate sensory modalities and, in some cases especially in children, reorganizes defective brain regions. Nonetheless, technological innovation has enabled the blind to read printed text (Kurzweil reader) and to convert printed material to vibrating pins on a fingerpad (Optacon scanner). Moreover, the evolution of brain areas that subserve mechanical skills underlies technological developments that abetted modern civilization. The rise of modern technology has freed the hands to think and the voice to proffer instruction. Indeed, sign language systems satisfy every criterion of a language in terms of generativity, syntax, semantics, and pragmatics. There is a syntax of movement as there is a syntax of speech and a logic of numbers as there is a physiology of logic as all skillful behavior involves the same aspects of sequence and seriation and packaged into a *generalized schema of action*. Kinesthetic thinking lies in orchestrating a sequence of activities; integrating intellective, emotional, and multisensory experiences; and then selecting and executing appropriate movement, action or activity.

Throwing, hitting, typing, writing, signing, singing, dancing, driving a car, and playing a musical instrument suggest that motor capacities are deeply involved with, and constitutive of, other intellective competencies. These activities partake of timing, force, selection, sequencing, orchestration, and integration that lie at the core of human intellective activity. The boundaries between perception, action, and cognition are porous. If human communication evolved from the capacity to recognize actions in early hominid populations to a mirror system for intentional gestured and spoken discourse, then thought, action, and perception are indissolubly tied. The organization of the brain and body is not top-down but organized at the level of the system that is dependent on local, distributed, and contextual constraints.

Thinking is an embodied activity. Although humans may be best characterized as symbol-using organisms, symbol use is structured by action and perceptual systems that occur in both natural and artefactual environments. Human consciousness may arise not just from some novel feature of human brains, but by way of the body's awareness of itself through it exteroceptive and proprioceptive senses. The body thus structures thought as much as cognition shapes bodily experiences.

Creative Thought

I elaborate below on the contribution of the body to creative thought building on my own theory of the development of early creative (metaphoric) abilities in infants and young children. I argue that creativity emerges out of a matrix of activities of the mind, brain, and body.

Perceptual Metaphor

It has been amply documented that, early on, very young children at the beginning of the third year of life are able to exploit the perceptual features of shape, color, texture, and size in making metaphorical associations in renaming single objects or pictures. For instance, a preschool child may refer to a plate of spaghetti as a "bunch of worms" or a vehicular caution sign as a "lemon ice cream cone." They can note these correspondences both in language as well as in match-to-sample tasks and similar nonverbal formats in which they must chose an analogous object from an array of pictorial or nonverbal choices.

Enactive Metaphor

Correspondingly, young children can make use of movement and motion information—that is, exploit enactive components of metaphor—in both categorizing moving objects and in making metaphorical associations. For example, a preschool child can note the metaphorical similarity between a spinning top and a dancing ballerina in a match-to-sample or similar nonverbal (filmed) format. Likewise, infants eighteen to twenty-four months can note the enactive similarity between a paper crayon cover that has slipped off and the putting on of an article of clothing, "I am putting your clothes on crayon."

Cross-Modal (Synesthetic) Metaphor

Infants and young children can also perceive likeness in different sensory modalities. An instance of such a synesthetic or cross-modal association would be a three-month-old infant that dishabituates or shows interest in the synchrony between a toy monkey and its complementary sound, a three-year old that applies the polar adjectives light/dark to an object (sandpaper) felt while blindfolded or a preschooler who indicates that red is a warm or hot color. The distinguished art historian, Ernest Gombrich, has long observed that such natural metaphors—bright sounds or cool colors—arise spontaneously in perception because they are part of our innate constitution.

Physiognomic Metaphor

During the very early preschool years rudimentary physiognomic experiences or the attribution of affective properties to visually perceived objects and other sensory experiences gain prominence. Examples include identifying the front grille of an automobile as a "smiling face" or a piece of music as "cheerful." In one large study, child and adult responses to physiognomically suggestive visual metaphor were examined in photographs. Preschoolers, 3 and 4 years of age, normal and high IQ children, 6 and 8 years of age, and adults were shown

10 photographs of cars, rocks, plants, and other inanimate objects and their responses categorized by type of metaphor. Preschool and 6-year-old children demonstrated significant levels of physiognomic responding, although high IQ, older children, and adults showed even higher levels of physiognomic responding presumably as a result of more advanced narrative abilities. All groups displayed high consistency in physiognomic object responses across photographs. It therefore appears that physiognomic metaphor is a robust phenomenon across the lifespan, originating in the very early childhood years, at least as early as the second half of the third year of life.

Evolutionary Considerations

Merlin Donald, a cognitive scientist, has suggested that in the evolution of primates, *metaphoric or integrative thought* originated in early primate episodic culture associated with the initial appearance of *perceptual metaphor* including *cross-modal associations*. For what reason would these abilities have evolved in early primates? Early primate cognition would have necessitated the ability to engage in complex event discrimination or *episodic cognition* and to react to complex patterns of stimuli based on earlier experiences or *episodic memory*—memory for events including associated emotions and the context, place, and time of the event—and such abilities would have been associated with both object and cross-modal perception. Comparable abilities, of course, are found in human infants and highlight the central role of neoteny in human development. Neoteny is the slowing down of development in the ontogeny of descendants such that a mental or physical ability in juveniles of ancestral forms is retained in adults. In this case, episodic cognition in humans is a cognitive legacy or an *archaic cognitive ability* inherited from earlier primates that proved useful to later evolving human primates in understanding conspecifics, their interactions, and events they initiated or encountered in the natural and social environment. This is called *episodic culture*.

Subsequently, *enactive metaphor* emerged in early hominid *mimetic cultures*. Professor Donald maintains that early hominid cognition would have required the ability to model behavior of conspecifics through vocal prosody, eye movement and gaze, facial expression, manual signs and gestures, rhythm, postural attitudes, and whole-body movements ("mimesis"). These early abilities would have combined primate event perception with a complex map of the body tied to self-initiated actions. That is, early hominids would have developed the ability to kinesthetically model perceptual events in the environment as well as the ability to recall those events, probably mediated by the *coeval* ("at the same time") emergence of mirror neurons in the human cortex. These capabilities would have evolved to facilitate toolmaking, fire and tool use, acclimation to seasonal changes in ecology and climate, participation in seasonal group hunting, and would have most likely also promoted social bonding through such actions as primitive pantomimic rituals. Not surprisingly, pretend action (enactive) metaphors appear in the speech and gestures of very young children.

More advanced metaphoric operations appeared, however, as human mythic cultures evolved approximately 250,000 years ago, precipitating complex language abilities and *integrative mythical thought*. Claude Levi-Strauss, the French anthropologist of the mid-20th century, argued that *mythical thought* arose from patterns in the mind that consisted of pairs of opposites (yes, *dichotomization*, again) such as strong/weak and good/evil. These patterns gave rise to a body of *stories* (syntagmatic or *narrative thought*) that gave meaning to the lives of early humans with regard to their own origins as well as that of other animals and plant species and the causes of natural forces and other phenomena including the characteristics of culture and its operative aspects. These mental abilities would have undergirded linguistic and complex gestural uses of metaphor as well as the emergence of simple analogies. Metaphoric thought in early human cultures—primarily *iconics* (movements which depict a bodily action), *beats* (movements which accent vocal speech), and *metaphoric gestures* (movements which depict a spatial relation)—would have enabled the integration of events across individual episodes and facilitated the ability to extract thematic content paving the way for symbolic languages and discourse, that is, complex paradigmatic (logical) and syntagmatic (narrative) thought. Iconic gestures, moreover, have been reported to emerge in human infants before 13 months of age and have been also reported in modern apes.

According to Professor Donald, more complex analogical reasoning including use of visual metaphor and other rhetorical devices were the outcome of modern human *theoretic (symbolic) cultures* that arose during the Upper Paleolithic period, 60,000 years ago or earlier. Under this evolutionary scenario, *visual thinking* emerged and extended metaphoric and iconic gesture to the verbal realm in which linguistic and other symbols were imputed to visual experiences as in "The road was a rocket of sunlight." In this sense, verbal metaphor and the conveyance of metaphoric relations through other symbols is secondary to action and perception in which metaphoric abilities originate.

In young children, visual metaphor emerges very early in child development. Thus, his timeline provides an evolutionary and biological basis for the evolutionary emergence of perceptual, enactive, and synesthetic (cross-modal) forms of metaphor in primate and hominid phylogeny.

With regard to physiognomic forms of metaphor, Stephen Mithen, a British anthropologist, contends that early metaphoric abilities arose in human evolution during the Middle/Upper Paleolithic period approximately 40,000 to 60,000 years ago due to increasing *cognitive fluidity* that allowed language to be metaphorically extended from primarily social to nonsocial uses. He suggests that anthropomorphic thought (inanimate objects given human characteristics) and totemic thought (animals represent humans groups), likely precursors of *physiognomic metaphor*, arose late in human evolution and had selective advantages including the ability to improve the predictive power and reliability of securing food and other essential material goods from animal and plant prey.

The emergence of rock art during this same period is helpful in understanding the evolutionary origins of anthropomorphism and totemism. Three types of signs are found in early rock art: *Pictograms*, that is, mythograms or representational figures of real or imaginary objects that consist of *zoomorphic* (animal-like) or *anthropomorphic* (human-like) features; *ideograms* or synthetic and repetitive signs (tree-shapes, phallic and vulvar signs, and groups of lines and dots); and *psychograms* or sharp graphic strokes that express perceptual experiences. These drawn and carved figures have clear metaphorical significance. For example, a particular sign may stand in place of individuals or tribal groups and function metonymically by association between these two things, as in the head of an animal standing in place of person. But it can also suggest an interplay between male and female attributes or qualities and function metaphorically by comparison between those two things, as in a vulvar sign connoting female sexual power.

Studies of cave art have come to similar conclusions suggesting that the origins of Paleolithic art originate in the body and in the complex interplay between human and animal signs. I have suggested elsewhere that such prehistoric signs have a strong affinity with physiognomic metaphor in which a percept is infused with an affective tone. This may partially explain the allure of such percepts in modern projective testing as the Rorschach Inkblot Test in which patients impute hidden or symbolic meaning to simple inkblots. Physiognomic metaphor, moreover, emerges very early in human development. The foregoing provides an evolutionary basis for the emergence of physiognomic forms of metaphor in later-evolving hominid populations.

Comparative Evidence

Comparative studies across species indicate that the core abilities underlying metaphor—such as the ability to form mental representations of perceived objects (to categorize and form concepts of objects; this round object over there is a tennis ball), integrate sensory and affective experiences of those objects (cross-modal transfer across sensory and affective modalities; this "perfume smells like bright sunshine"), and form relational concepts (a sister's relationship to her daughter's offspring)—may be shared with other birds and mammals. For instance, in all investigated primate species there is evidence for cross-modal transfer between the visual and haptic modalities. Chimpanzees have shown the ability to transfer between the visual and gustatory modalities, between the visual and auditory modalities, as well as across some classes of symbolic stimuli. Bonobos demonstrate the latter ability as I discussed earlier. Bushbabies (prosimians) are able to transfer learned responses from the visual to the auditory modality. Chimpanzees and other apes and monkeys (rhesus macaques, squirrel and capuchin monkeys) can form relational categories including the understanding of *identity* (which two objects are identical? or A = A'), *oddity* (which object is the odd one? or B is not A or A'), and *same/difference* (which two objects are different? or A is not B) problems in match-to-sample like tasks. Chimpanzees are also capable of constructing multiple groupings of objects in a spontaneous manner and even engage in insight in solving some kinds of physical problems. Not surprisingly, elephants reportedly do. In essence, some non-human animals are able to make an abductive inference and arrive at the solution to a problem. For instance, a young male elephant moved a large cube to step on to acquire some food that was out of reach. Psittacines (parrots) demonstrate Piagetian stage 6 object permanence as we saw in an earlier chapter, form categorical classes of objects (all these objects over here are red and square), label and enumerate natural and unnatural objects, fashion concepts of same/different, and display some comprehension of relative size.

The foregoing indicates that the core abilities underlying metaphor—the ability to form mental representations of objects, integrate sensory and affective experiences, and form relational concepts—are widely distributed in the animal kingdom presumably because they have evolved to solve similar adaptive problems in the natural world or have emerged for complex social purposes that have driven the application of these abilities beyond their immediate Darwinian adaptations. For example, *sentinel behavior* or the need to protect groups of conspecifics from predators as in the highly social meerkat (a small African burrowing carnivore of the mongoose family), may have been part and parcel of the ability to categorize objects in the natural world, including conspecifics, and for doing so in a flexible manner using available information in the environment. Such a *universal toolkit* or the evolution of *domain-general abilities* that are common across species—the ability to recognize objects, enumerate, navigate, and acquire knowledge—could be the basis of a species' ability to relate percepts, affects, and concepts across disparate domains of experience, that is, create the conditions for the evolution of the ability to make metaphoric discriminations. The ability to combine simpler elements into more complex structures appears to be a fundamental property of biological systems.

The developmental (ontogenetic), evolutionary (phylogenetic), and comparative evidence, thus, indicate that perceptual, enactive, synesthetic, and physiognomic metaphoric abilities emerge very early in life, are biologically grounded in the mind-brain-body (MBB) and form the ability to perceive likeness across certain domains and may further lie at the core of creative thought. Only secondarily, is the ability to perceive likeness or nonliteral similarity captured in different symbol systems such as gestures, words, pictures, numbers, musical sounds, dance notation, and so on.

Bodily Basis of Metaphor

Motor and Nonmotor Areas of the Brain Overlap

The evidence for the bodily or embodied basis of metaphor is fourfold. In the first instance, many of the so-called motor structures in the brain (cerebellum, basal ganglia, and premotor and motor cortices) play a significant role in cognition. For example, the acquisition in children of two-word utterances during the second half of the second year of life is revealing. Roger Brown, a psycholinguist, has documented that eleven types of sensorimotor experience are commonly manifested in infants' initial two-word combinations. For instance, the relationship of agent performing an action is established during the sensorimotor period and then later mapped onto early language as in "Johnny fall." These eleven fundamental types of bodily-kinesthetic knowledge include forms of reference (naming, recurrence, and nonexistence) as in the linguistic utterance, "There doggie";

various sets of relations (agent-action, agent-object, action-object, action-location, and object-location) as in the utterance, "Book table"; possessor and possession ("My ball"); object attribution ("Little story"); and the category of demonstrative entity in which object specifications are applied to objects ("That ball"). These abilities are made possible because brain areas for motor and cognitive functions overlap and are interdependent and in acquiring conceptual knowledge the two are inextricably intertwined. Thus, enactive or action metaphors in which a spinning top is likened to a dancing bear, emerge in early childhood as a consequence of the acquisition of sensorimotor knowledge acquired during the first two years of life.

Mirror Neurons

A second piece of evidence is that there is a mirror system for both action execution and action recognition in the primate cortex suggesting the close association between intended movements carried out by an actor and observed movement performed by a viewer. This system is underwritten by *mirror neurons* in the premotor cortex that link externally observed events (manual signs and gestures, postural attitudes, facial expression, eye movements and gaze, vocal prosody, whole-body movements, and rhythm) to internally generated motor actions and establishes a relationship between gesture and intentional communication facilitating both the establishment of intentional mental states and the ability to read them in others. This mirror system would enable primates to model external events in self-initiated motor behavior ("mimetic skills") and such a capacity would underlie intentional nonverbal communication and eventually speech. Some observers argue that this mirror system would also underwrite *sensorimotor synesthesia* or the mimicking of auditory rhythm by rhythmic movements. Taken as a whole, forms of motor-to-motor mappings or *synkinesia* would be mediated by mirror neurons and underlie enactive metaphor.

This mirror system is also generative. It has been argued that there is a general-purpose sequencing module in the brain that parses motor action into its components, recombines these components, and these recombined components underlie both movement and speech. Spoken language and movement such as gesture would draw on the same underlying brain systems. Thus, it has been hypothesized that the evolutionary basis for vocal languages originates in gesture. Contemporary studies of signed languages lend some support to this view. For instance, sign language expresses metaphoric meaning through the modification of sign by way of the overlapping, blending, and substitution of signed gestures. Ironic and metaphoric modes of communication are quite common in deaf communities including aesthetic uses of sign as in deaf theatre or sign poetry. Deaf children, too, develop the capacity to produce figurative sign language through use of pantomime, ritualized movement, sign modification, and by adding iconic, visuospatial detail. Similarly, research has shown that blind subjects are able to depict motion metaphorically in drawings and will draw, for example, curved spokes on paper to depict the movement of a wheel. Not surprisingly, both deaf and blind subjects, whether children or adults, make use of enactive and perceptual metaphor in acquiring sign language, Braille or spoken language.

Cross-Modal Perception

In the third instance, there is evidence for cross-modal perception in infants and very young children that arises from the close association between sensory, perceptual, and cognitive processes that are intimately tied to the body. For instance, one study demonstrated that 4-, 6-, and 8-month olds can discriminate audible, visible, and bimodal components of isolated syllables. Each infant watched and listened to a person mouthing and uttering a syllable and was then assessed in their ability to detect any changes in the audible, visible or combined (audible-visible) properties of the syllable. By 6 months, infants picked up on the visible and audible properties of the syllables but by 8 months infants grasp the interplay between the audible and visual properties or *bimodal components* of syllables. Nonetheless, some evidence suggests that while very young infants appear to match intersensory stimuli based on equivalence of the intensity of the stimulus, older infants and very young children match cross-modal stimuli based on complete multisensory integration that includes the concatenation of sensory, perceptual, and cognitive information. A similar view stresses that early cross-modal abilities are only apparent in early infancy or what's called *neonatal synesthesia* based on cross-modal confusion and the lack of cortical specialization. True cross-modal transfer appears to kick in before the end of the first year or sooner and is based on connections among fully differentiated sensory areas.

Representation of the Body Schema

As I discussed in Chapter 2, there is evidence for a distributed neural network, including the somatosensory cortex, limbic system, and cortical regions central to object- and self-recognition that form the neural basis for corporeal awareness, that is, one's body or body schema. One's sense of self and personal identity are constructed from one's body, namely, body-centered memories that are modulated over time by changes to one's body schema or plan. It is plausible that the ability to perceive and exploit early visual-affective and other sensory-affective relations that rely on physiognomic characteristics—the outward appearance or character of an individual—is a result of the emergence of this distributed neural network very early in development as well as evidence that this network forms the basis for bodily awareness. It has been suggested that in understanding others a cross-modal system has evolved that binds self and other principally through imitation. Such a system appears to be innate and allows even very young infants as early as six weeks of age to imitate others based on their own primitive body scheme.

Multisensory and Multimodal Integration

Research on the brain basis of multisensory integration indicates that the superior colliculus (SC), the anterior ectosylvian cortex, and the rostral portion of the lateral suprasylvian sulcus—structures located in the midbrain and frontal cortices—are comprised of multisensory neurons that facilitate multisensory integration of visual, auditory, somatosensory, and motor experiences in individuals. In the primate cortex, the integration of vision, touch, proprioception, and motor feedback occurs in area 5 of the parietal cortex, that is, the intraparietal sulcus (IPS), and underlies representation of the body plan. It may also provide the neural basis of physiognomic metaphor along with ventral visual areas of the cortex (V4 for facial perception), the amygdala, and related limbic areas for the perception of emotion.

Area V5 in the superior temporal sulcus (STS) appears to directly serve the detection of biological motion and would seem to be a likely candidate for the neural basis of enactive metaphor as well as auditory and rhythmic synchronicity. Rhythm is mediated by the right posterior superior temporal gyrus and functions in the human perception of music, among other things. The STS appears to function as a *convergence zone* for integration of the dorsal "where" and the ventral "what" visual pathways as well as having important connections to the amygdala and orbitofrontal cortex (OFC) for the interpretation of stimuli. The inferior temporal cortex houses memory for shape along with areas for color (V4) and dynamic form (V3) in the occipital cortex and would certainly be likely candidates for the neural basis of perceptual metaphor.

The superior colliculus (SC) and an adjacent area in the cortex, the insula, seem to be implicated in the amalgamation of crossmodal information based on temporal correspondence. The inferior parietal sulcus (IPS) is likely involved in the assemblage of cross-modal information based on spatial correspondence particularly in the anterior region of the IPS. Both of these and related structures would be probable candidates for synesthetic or cross-modal metaphor. Similarly, the superior temporal sulcus (STS) has been posited to coordinate audiovisual information in speech whereas the insula and claustrum (CS), which are situated next to each other in the midbrain, appear to integrate the visual and tactile modalities. The rhinal cortex in the medial temporal region is involved in the storage of multimodal information, the claustrum in their cortical synchronization, especially cortical synchronization of sensory-specific cortices, and the superior colliculus (SC) playing a major role in the mediation of orienting and attentive behaviors.

Some observers maintain that both cortical and subcortical structures are involved in multisensory and multimodal integration at several brain sites. The cortical sites include the superior temporal sulcus (STS), the intraparietal sulcus (IPS), and areas in the frontal and parietal lobes. They reason that the frontal cortex is implicated in the initial learning of newly acquired multimodal associations whereas the STS appears to be involved in the subsequent learning of amodal representations. The IPS facilitates the multimodal integration of spatial coordinate information as well as the integration of vision and touch.

The subcortical sites include the supergeniculate and medial pulvinar nuclei of the thalamus, which connect to the superior colliculus and visual and auditory areas. On the other hand, the nucleus reticularis thalami (NRT) is responsive to polymodal cortices. When inhibited by these cortices it increases cortical arousal and potentiates the mesencephalic reticulating system (MRF) enabling the organism to be more responsive to novel and biologically significant stimuli, thus enhancing the brain's ability to pick out relevant percepts based on these fundamental types of metaphoric processes.

On that score, some neuroscientists contend that multisensory integration evolved in primates and other species to increase the probability of stimulus detection, particularly stimulus novelty and significance, heighten the ability of organisms to simultaneously monitor multiple environmental cues (yes, *multitasking*, again), facilitate speed and accuracy of identification, as well as the ability to detect misinformation.

Creative thought in human primates would thus represent an exaptation of multimodal abilities that originally evolved for the servicing of more mundane activities essential to survival in other species. Neuroscientists have made similar claims for the evolutionary role of the angular gyrus in synesthetic or cross-modal metaphor. That is, the angular gyrus (BA39 in the anterolateral parietal lobe) interleaves the sensory modalities (audition, vision, and touch) creating the preconditions for abstract thought. This aspect of cognition involves extracting sensory information—such as size, shape, and texture—and formulating a more complex percept or abstract property such as "jaggedness."

What happens when thought goes awry? Let's look at the pathologies of the body that disrupt thought.

Notes

1. The quote, "Physical intelligence includes ..." is from Stephen Jay Gould, The brain of brawn, 2000.
2. The quote, "The body is shaped ..." is from Martha Graham, I am a dancer, 1952.

Bibliography

Anati, E. (1994). *World rock art: The primordial language*. Valcamonica, Italy: Edizioni Del.

Arnheim, R. (1998, August-September). Rudolf Arnheim: "Die Intelligenz des Sehens," *Neue Bildende Kunst*, pp. 56–62.

Barten, S. S. (1979). Development of gesture. In N. R. Smith & M. B. Franklin (Eds.), *Symbolic functioning in childhood* (pp. 139–151). Hillsdale, NJ: Lawrence Erlbaum.

Bebko, J. M., Burke, L., Craven, J., & Sarlo, N. (1992). The importance of motor activity in sensorimotor development: A perspective from children with physical handicaps. *Human Development*, *35*(4), 226–240.

Berlucchi, G., & Aglioti, S. (1997). The body in the brain: Neural bases of corporeal awareness. *Trends in Neurosciences*, *20*(12), 560–564.

Brown, D. E. (1991). *Human universals*. Boston, MA: McGraw-Hill.

Brown, R. (1973). *A first language: The early stages*. Cambridge, MA: Harvard University Press.

Butterworth, B. (1999). A head for figures. *Science*, *284*(5416), 928–929.

Butterworth, B., Gallistel, C. R., & Vallortigara, G. (2017, Nov. 7). The origins of numerical abilities. *Philosophical Transactions of the Royal Society B*, *373*, 20160507. Retrieved from http://dx.doi.org/10.1098/rstb.2016.0507

Dehaene, S. (2011). *The number sense: How the mind creates mathematics* (Rev. & Expanded Ed.). Oxford, UK: Oxford University Press.

Dehaene, S., Spelke, E., Pinel, P., Stanescu, R., & Tsivkin, S. (1999). Sources of mathematical thinking: Behavioral and brain-imaging evidence. *Science*, *284*(5416), 970–974.

Donald, M. (1991). *Origins of the modern mind: Three stages in the evolution of culture and cognition*. Cambridge, MA: Harvard University Press.

Gardner, H. (1993). *Frames of mind: The theory of multiple intelligences*. (Rev. ed.). New York, NY: Basic Books.

Gilarranz, L. J., Rayfield, B., Liñán-Cembrano, G., Bascompte, J., Gonzalez, A. (2017). Effects of network modularity on the spread of perturbation impact in experimental metapopulations. *Science*, *357*(6347), 199–201.

Goldin-Meadow, S. (1997). When gestures and words speak differently. *Current Directions in Psychological Science*, 6(5), 138–143.

Gould, S. J. (2000, June 25). The brain of brawn. *The New York Times*, p. A20. Retrieved from https://www.nytimes.com/2000/06/25/opinion/the-brain-of-brawn.html.

Graham, M. (1952/1998). I am a dancer. In A. Carter (Ed.), *The Routledge dance studies reader* (pp. 66–71). New York, NY: Routledge.

Graziano, A. B., Peterson, M., & Shaw, G. L. (1999). Enhanced learning of proportional math through music training and spatial-temporal training. *Neurological Research*, *21*(2), 139–152.

Hadamard, J. (1996). *The mathematician's mind: The psychology of invention in the mathematical field*. Princeton, NJ: Princeton University Press. (Original work published 1945).

Hauser, M. D. (2000). *Wild minds: What animals really think*. New York, NY: Henry Holt.

Jones, S. (2005). The role of mirror neurons in imitation. In S. Hurley & N. Chater (Eds.), *Perspectives on imitation: From neuroscience to social science*. Vol. 1. Mechanisms of imitation and imitation in animals (pp. 205–210). Cambridge, MA: MIT Press.

Kennedy, J. M. (1997). How the blind draw. *Scientific American*, *276*(1), 76–81.

Kimura, D. (2000). *Sex and cognition*. Cambridge, MA: MIT Press.

Klima, E. S., & Bellugi, U. (1980). Wit and poetry in American sign language. In W. C. Stokoe (Ed.), *Sign and culture* (pp. 105–132). Silver Spring, MD: Linstok Press.

Lakoff, G., & Nunez, R. E. (1997). The metaphorical structure of mathematics: Sketching out cognitive foundations for a mind-based mathematics. In L. D. English (Ed.), *Mathematical reasoning: Analogies, metaphors, and images* (pp. 21–89). Mahwah, NJ: Lawrence Erlbaum.

Lewkowicz, D. J. (1994). Development of intersensory perception in human infants. In D. J. Lewkowicz & R. Lickliter (Eds.), *The development of intersensory perception: Comparative perspectives* (pp. 165–203). Hillsdale, NJ: Lawrence Erlbaum.

Lewkowicz, D. J. (2000). Infants' perception of the audible, visible, and bimodal characteristics of multimodal syllables. *Child Development*, *71*(5), 1241–1257.

Mak, B., & Vera, A. (1999). The role of motion in children's categorization of objects. *Cognition*, *71*(1), B11–B21.

Marschark, M., Everhart, V. S., Martin, J., & West, S. A. (1987). Identifying linguistic creativity in deaf and hearing children. *Metaphor and Symbolic Activity*, *2*(4), 281–306.

Maurer, D. (1997). Neonatal synesthesia: Implications for the processing of speech and faces. In S. Baron-Cohen & J. E. Harrison (Eds.), *Synaesthesia: Classic and contemporary readings* (pp. 224–242). Oxford, UK: Blackwell.

Maurer, D., & Mondloch, C. J. (2005). Neonatal synesthesia: A reevaluation. In L. C. Robertson & N. Sagiv (Eds.), *Synesthesia: Perspectives from cognitive neuroscience* (pp. 193–213). Oxford, NY: Oxford University Press.

Mithen, S. (1996). *The prehistory of the mind: A search for the origins of art, science, and religion.* London: Thames & Hudson.

Mithen, S. (2006). *The singing neanderthals: The origins of music, language, mind, and body.* Cambridge, MA: Harvard University Press.

Nicoladis, E., Mayberry, R. I., & Genesee, F. (1999). Gesture and bilingual development. *Developmental Psychology, 35*(2), 514–526.

Ramachandran, V. S., & Hubbard, E. M. (2005). The emergence of the human mind: Some clues from synesthesia. In L. C. Robertson & N. Sagiv (Eds.), *Synesthesia: Perspectives from cognitive neuroscience* (pp. 147–190). Oxford, NY: Oxford University Press.

Ramachandran, V. S., Rogers-Ramachandran, D., & Steward, M. (1992). Perceptual correlates of massive cortical reorganization. *Science, 258*, 1159–1160.

Sacks, O. (1988, June 2). The revolution of the deaf. *The New York Review of Books*, 23–28. Retrieved from https://www.nybooks.com/articles/1988/06/02/the-revolution-of-the-deaf/

Sacks, O. (1990). *Seeing voices: A journey into the world of the deaf.* New York, NY: HarperCollins.

Schwartz, D. L. (1999). Physical imagery: Kinematic versus dynamic models. *Cognitive Psychology, 38*(3), 433–464.

Seitz, J. A. (1997). Metaphor, symbolic play, and logical thought in early childhood. *Genetic, Social, and General Psychology Monographs, 123*(4), 373–391.

Seitz, J. A. (1997). The development of metaphoric understanding: Implications for a theory of creativity. *Creativity Research Journal, 10*(4), 347–353.

Seitz, J. A. (1998). Nonverbal metaphor: A review of theories and evidence. *Genetic, Social, and General Psychology Monographs, 124*(1), 121–143.

Seitz, J. A. (1999, June 4). Selfish sentinels in cooperative mammals [Letter to the editor: Response to T. H. Clutton-Brock]. *Science, 284*(5420), 1640–1644.

Seitz, J. A. (2005). The neural, evolutionary, developmental, and bodily basis of metaphor. *New Ideas in Psychology: An International Journal of Innovative Theory in Psychology, 23*, 74–95

Seitz, J. A., & Beilin, H. (1987). The development of comprehension of physiognomic metaphor in photographs. *British Journal of Developmental Psychology*, 5(4), 321–331.

Stein, B. E., & Meredith, M. A. (1993). *The merging of the senses.* Cambridge, MA: MIT Press.

Sternberg, M. L. A. (1999). *The American Sign Language Dictionary* [CD-ROM]. Seattle: Multimedia 2000.

White, R. (1994). Les archives paléolithiques. *La Recherche, 25*, 746–750.

Winner, E. (1995). Introduction. 10th anniversary special issue: Developmental perspectives on metaphor. *Metaphor and Symbolic Activity, 10*(4), 247–253.

Winner, E., McCarthy, M., & Gardner, H. (1980). The ontogenesis of metaphor. In R. P. Honeck & R. R. Hoffman (Eds.), *Cognition and figurative language* (pp. 341–361). Hillsdale, NJ: Lawrence Erlbaum.

Winner, E., McCarthy, M., Kleinman, S., & Gardner, H. (1979). First metaphors. In D. Wolf & H. Gardner (Eds.), *Early symbolization: New directions for child development*, (vol. 3, pp. 29–41). San Francisco, CA: Jossey-Bass.

· 5 ·

DISRUPTION OF THOUGHT IN PATHOLOGIES OF THE BODY

The disruption of thought in the pathologies of the body such as stuttering, Parkinson's and Huntington's diseases, and autism comprise the current chapter. Some of these diseases I have diagnosed or treated first hand, whereas others such as Kallman's syndrome, I have relied on case histories and, of course, the scientific and medical literature.

Stuttering

Stuttering, which is also known as stammering, is a problem in language fluency. It may involve repeating words or phrases or other language sounds (syllables) unintentionally, unusual prolongation of speech sounds that are produced without complete closure of the vocal tract ("yessssssterddday") or the tendency to hesitate in one's speech or produce long pauses between words or phrases. From the standpoint of a mind-brain-body (MBB), stuttering impairs an individual's competence in concatenating the various muscular movements that underlie spoken fluency.

There are three major theories of stuttering and in the first two, genetic factors are implicated. In the first instance, there is an inability of the nervous system to control the speech articulatory apparatus (larynx, tongue, the upper and lower lip, the hard and soft palate, pharynx, and epiglottis) resulting in a condition called *neurogenic or acquired stuttering*. Neural factors are the cause. Indeed, malfunctioning of the basal ganglia, that extremely important subcortical structure we previously discussed, appears to be the culprit. In the second instance, the acquisition of grammar during early childhood may precipitate stuttering and this condition is called *developmental stuttering*. Some have suggested that this is actually not a grammatical problem but a word-finding problem as the child appears to have issues in retrieving the right word from memory. The general consensus, however, is that the first two are variants of a disruption of the basal ganglia-thalamocortical motor circuit by way of the putamen to the supplementary motor area (SMA). The putamen is one of the three areas of the striatum in the basal ganglia and receives input primarily from the frontal and sensorimotor cortices and outputs information directly to the motor cortices including the primary (MC), premotor (PMC), and supplementary motor areas (SMA). Other cortical areas may also be involved.

In the third instance, the initiation and maintenance of stuttering can be influenced by the social context as well as the individual's emotional reactions to others as in the feeling of frustration in mucking up a task. On the one hand, dysfluencies tend to decrease when the stutterer is speaking alone or attempts to sing or hum the words. On the other hand, emotional reactions such as anxiety, shame or embarrassment can increase stuttering. As a result, these symptoms as a whole are known as *psychogenic stuttering*. Stuttering, thus, may sometimes have a psychological and/or social cause.

Interestingly, the prevalence rate is in a three-to-one (3:1) ratio of males to females and might be expected if the emergence of speech is affected by the cerebral organization of the human brain. This organization differs in males and females and may account for the statistical advantage that is found in females for development of speech. Standardized tests that measure verbal intelligence in women and men don't show that women have more verbal intellectual capacity, but they do seem to shine in grammatical usage, verbal fluency, verbal memory, and spelling. They may be better at recalling numbers or numerical memory but not incidental memory or things they heard but weren't paying attention. The question then becomes what might account for these differences in women and men? For one thing, most speech functions in men are typically housed or lateralized to the left hemisphere of the brain whereas in women they are more symmetrically distributed to both the left and right hemispheres. Verbal fluency and vocabulary, but not verbal memory, tend to be bilaterally organized in women. As a result, if there is damage to the speech circuits in the left hemisphere as in stuttering, men may be more greatly affected than women.

Singing words appears to mitigate dysfluency in stutterers. It has been suggested that irregularities in timing by the basal ganglia-thalamocortical loop may account for the triumvirate of motor-speech behaviors that greatly abet the attenuation of stuttering: Singing, chorus speech, and the rhythm effect. Speaking at the pace of a metronome—known as the *rhythm effect*—can reduce stuttering by compensating for inappropriate timing cues originating in the basal ganglia. So, can reading in unison with others or what's called *chorus speech*. Singing also works. But, why? Music provides a separate mechanism for generating internal timing cues when there is aberrant output from the basal ganglia and this mechanism most likely occurs by way of the premotor cortex (PMC). There is also some evidence that modified auditory feedback can improve dysfluent speech in stutterers. What's interesting from the perspective of the mind-brain-body (MBB) is the contribution of the body to the functioning of the intellect. By adjusting one's larynx and speaking at a regular pace the brain's timing is re-calibrated. And, by reading or singing with others in a social context one is able to accomplish a similar feat. It is not that the body gives mere feedback to the brain but is very constitutive of thought, or in this case, a specific component of language, that is, timing. Of course, we can program a computer to speak but nothing inherent in its output device modifies the syntax, pragmatics or semantics of the sentence. Or, at least not yet. And when that happens the computer's physicality will be an inherent component of the computer's operation.

Autism (Autistic Spectrum Disorders—ASD)[1]

Autism, a rare but devastating disorder of early childhood, results in profound social impairment, wide-ranging deficits in verbal and nonverbal communication, and extensive behavioral stereotypies in infants and young children that collectively interfere with social and emotional development. All major neuropathological theories of autism place its genesis in abnormalities during the prenatal and/or perinatal periods, that is, a congenital birth defect that frequently goes undiagnosed until the early childhood years.

For our purposes, autism is a profound disorder of the mind-brain-body (MBB) that results in pathological alterations of behavior including the presence of behavioral rituals that involve, among other things, repeating the same behavioral task over and over again, as well as an insistence on sameness in interaction with objects and people. The nature of the social deficits, however, has increasingly taken front stage in studies of autism. Newer theories of intellective processes indicate that social intelligence consists of two essential components. Intrapersonal abilities or emotional intelligence include one's affective range and intensity, the ability to discriminate among the emotions, the ability to label them appropriately, and the ability to use them to guide one's behavior. Interpersonal abilities or social intelligence proper include the ability to decode feelings, intentions, and motivations in others; recognize characteristics among people (age, gender, and ethnicity); and influence others to behave in desired ways. Autistic children's social impairment appears to involve deficits in both intrapersonal and interpersonal abilities that impact on verbal and nonverbal communication.

In autism, despite the profound emotional impairment found in autistic children, some researchers have claimed that there is instead a central cognitive deficit in social understanding such as imputing beliefs to others. However, these theoretical claims fail to account for the emotional nature of the social deficits. Specifically, they have ignored autistic children's failure to use emotions as a source of knowledge. Their claims neglect to take account of numerous reports and frequent observations that there are profound deficits in the decoding, range, expression, and symbolic labeling of the emotions as well as their interpersonal elaboration as in their lack of empathic understanding or awareness of the feelings of others. As regards intrapersonal abilities, autistic children show a preference for pairing facial expressions with non-affective stimuli and using the lower half of the face in recognition of peers unlike normal children who use the upper half of the face. They are also poor at matching various graphic representations of emotions with facial expressions. As regards interpersonal abilities, autistic children are poor in understanding humor, in detecting social deception in picture stories, in differentiating the social categories of age and gender, and in role-taking tasks that require empathic understanding.

The social deficits, moreover, appear to affect the class of symbolic skills that autistic children acquire. That is, autistic children's social deficits appear to interfere with the acquisition of symbolic skills dependent on social processes such as communicative language, symbolic play, and empathy. For example, the word "apple" serves as a label or symbol for the actual physical object. So, a child may use a broomstick to represent a horse in play or offer to another child a toy as symbolic of his feelings towards that child. These symbolic skills arise largely as a result of social interaction with others as attested to by the literature on language development, development of symbolic play, and growth of empathic awareness in young children.

Other symbolic skills acquired in the early childhood years are largely independent of social processes such as visual-spatial and number skills, attainment of object permanence, and cross-modal processing of stimuli and appear not be compromised to the same extent in autistic children. The development of these skills does not require extensive emotional processing by areas of the brain that subserve social interaction. As a result, autistic children demonstrate enhanced performance on some cognitive tasks in contrast to poorer performance on many socially influenced tasks.

The central pathology in autism, then, is a defect in the ability to discriminate and symbolically elaborate upon the emotions and/or a defect in the ability to attribute and decode feelings in others, that is, impaired empathic understanding. Both of these processes are intimately tied up with the body and thus represent a profound disorder of the MBB. As an aside, certain neurodevelopmental disorders of childhood such as Asperger's syndrome, Rett syndrome, and childhood disintegrative disorder are now grouped together as spectrum disorders of autism. Indeed, prospective studies of 106 human infants with familial risk for an autistic spectrum disorder (ASD) are illuminating. In this case, these infants had another family member who had been diagnosed with ASD making the infant more vulnerable to getting the disorder, but it's a prospective study so they haven't been yet diagnosed with autism. But they did show *hyperexpansion* or overexpansion of cortical surface area at 6–12 months, and some children developed abnormal increases in brain volume at 12–24 months, as 15 were later diagnosed with autism by the end of the second year of life when autistic behaviors first emerge. Hyperexpansion of the cortical mantle and abnormal increases in brain volume are physical *markers* in the brain for ASD. If they emerge in the first two years of life, a child may be vulnerable to a diagnosis of autism.

Both intrapersonal and interpersonal abilities quintessentially operate through the body such as expressing or decoding facial expressions, gesture, and body posture; maintaining interpersonal distance; and gauging gender, age, sexual orientation, and ethnicity. Access to one's emotions, the ability to symbolically elaborate on one's feelings, the skill in using cognitive-affective states to guide one's behavior, and the capacity to read intentions in, and influence others, constitutes a working definition of personal identity and one's sense of self. However, personal identity is not solely mind-based. As we discussed earlier, the body influences personality by tagging affective states with somatic markers. If you remember, a somatic marker is a visceral or non-visceral sensation that marks a cognitive-affective state. It is acquired through internal biological preferences ("I love dark chocolate") as well as external cultural prescriptions ("I like celebrating my birthday with my two best friends"). Basic, bodily regulatory systems thus influence conscious cognitive processes.

One could add that subjectivity itself arises from changes that take place in one's bodily states. That is, subjective awareness of self, arises by way of the body's awareness of itself through interoceptive (internal chemical senses), proprioceptive (sensors in our muscles and joints), and exteroceptive senses (the five external sensory modalities). From the study of the brain's prefrontal cortices, three sets of neural structures appear to underlie this sense of self: Our sensory cortices, our sensorimotor cortices, and subcortical structures including the cerebellum, basal ganglia, and thalamus. Your sensory cortices record the outside world of people, objects, and things; your sensorimotor cortices record the internal world of your body primarily through proprioception; and your subcortical structures create a link between these two worlds of outside and inside. Since these linked structures are nonverbal and constitute bodily-based functions, they might also exist in infrahuman primates and other animals. Your African grey parrot has a personality. Your Siamese cat has character. Let's step back for a moment from our discussion of autism and examine the traditional mind-based self.

While traditional theories of selfhood have posited that personality is wholly mind-based, new views of the brain and the role of sensorimotor capacities in cognition and development, suggest that one's sense of self and personal identity are constructed from body-centered representations that are modulated by changes to one's body schema. Evidence from both case studies of self-image in relation to cultural views of beauty and the breakdown of personal identity after brain damage and as a result of mental illness, tend to support this view: The bodily basis of self and personal identity.

In the first case study, I present a 40-year-old female with a personal history of self-preoccupation with her body modulated by her sense of identity and changes in her body with age that were tied to her sense of identity.

Case Study #1

The subject (L. S.) is a single, female, 40-year-old, advanced degree, financial sector employment, presently unemployed, who lives in a large urban area.

L. S. is 5'6" tall, approximately 105 pounds, slender and small-boned. She comes from a family of four daughters, two or which are married and one of whom is divorced. L. S. describes herself as a late-bloomer with secondary sex characteristics developing later than normal. She began creative movement classes in fourth grade and by fifth grade had started ballet. She continued ballet until eighth grade and then took it up again after college. She grew up in the suburbs of a large urban area and had many friends until college where she found herself unaccepted and isolated socially. In college, she experienced problems in focusing her attention on studying. Instead, she began to focus her attention on her body, eating, and exercise. Although she received a fair amount of romantic attention from men, in her junior year, she weighed about 100 pounds and felt "unsatisfied with the way she looked." L. S. describes herself as shy as a child and hypersensitive since childhood. However, she is able to speak easily about her feelings and expresses them fluidly through dance. In other words, she has a good sense of who she is but her interactions with others are problematic.

Growing up, she reports feeling particularly uncomfortable with games with peers that involved touch such as writing on each other's backs in the family pool. Her parents and family stressed career and social status and criticized her for being "too thin or not eating enough." On the other hand, many but not all of her peers emphasized being physically attractive and appealing to the opposite gender. Currently, L. S. describes herself as being not being proud but shameful of her body: "I don't deserve food or sex because I am such a miserable person." She claims her "body feels dirty" and a "used good." She believes that pregnancy will lead to permanent damage to her body. L. S. has experienced premenstrual syndrome starting in her mid-thirties that makes her feel tense, emotionally labile, irritable, consumed by frequent headaches, difficulty relaxing, and experiences food cravings. She compensates for her change in her sense of self through use of alcohol in the evenings to relax—although recently curtailed—and physical exercise including running, swimming, floor barre classes, weight training, and yoga. L. S. spent a year or so in therapy with a psychiatrist who placed her on Zoloft, a serotonin agonist. However, she reports only moderate changes in the way she feels about herself. She reports dressing in an understated way as well as making little use of cosmetics or coiffed hair.

L. S. is overwhelmed by negative feelings about her body. Her life and personality are deeply affected. She was hypersensitive to others and experienced social isolation and touch aversion but found solace in aesthetic movement.

She had premenstrual syndrome with accompanying stress and used alcohol to relieve anxiety rejecting pleasurable experiences of the body such as sex and food. She immersed herself in movement, action, and the activity of sports, dance, and yoga, which helped her focus her attention and experience her mind-brain-body (MBB) more fully.

Gilles de la Tourette Syndrome

A gripping disorder, Gilles de la Tourette Syndrome is characterized by a number of defining symptoms: Grunting vocalizations that are incessant, facial and bodily tics that are unrelenting, a predisposition to utter profanities, and an overwhelming desire to scrutinize one's surroundings. As a result, Touretters often develop heightened social awareness as a means of dealing with uncomfortable social circumstances. Nonetheless, in some cultures, such as the Zulus of South Africa, Indiki or "jerk illness" as it is called, is considered sacred and those afflicted are often promoted to tribal healers or chiefs.

From the perspective of the thinking body, conversation with Touretters is punctuated by gesticulation and gesture as if it were kind of an activity-based punctuation in which movement of the body and motor speech acts are intertwined. As a consequence, habit reversal therapy, combining both behavioral and cognitive interventions as well as medication such as Pimozide (Orap)—a dopaminergic blocking agent—are treatments of choice. These treatments appear to ameliorate the information-processing demands of the caudate nucleus in the basal ganglia along with its interaction with the frontal lobes in the cerebral cortex and, thus, undermine the compulsive aspects of Tourette's Syndrome.

Obsessive-Compulsive Behavior (OCD)

The ability to shift from one activity (doing one's taxes) to another (answering the phone) involves inhibiting the first activity to pursue another course of action and results in the production of novel behaviors or novel sequences of behavior. Repetitive stereotyped movement patterns—obsessive-compulsive behavior and Gilles de Tourette syndrome—indicate the malfunctioning of this system. You're literally stuck in a cliched set of actions without recourse from your manual tyranny: Addictive behaviors (gambling, some forms of market investing, drugs, and sex) are one common example, but you also see individuals engaged in stereotypical thinking about politics, about relationships, about how to carry out a particular procedure at work, about what car to purchase, where to live, what to wear, and so on.

In psychiatry, these are grouped together as OCD spectrum disorders: Gambling, paraphilia (sexual fetishes), body dysmorphic disorder (thinking you are fat when you're thin), trichotillomania (constant hair-pulling), hypochondriasis, somatization disorder (frequent psychosomatic complaints), Gilles de Tourette syndrome, autism and Asperger's syndrome, kleptomania, impulse control disorders, obsessive-compulsive personality disorder, bulimia, and anorexia nervosa.

The cortex, namely, the orbitomedial frontal cortex; the subcortex, particularly the basal ganglia; and the body work together in communion with the social and physical environment to accomplish commonplace intellective tasks. The orbitomedial frontal cortex is involved in inhibiting, often, socially-inappropriate behaviors and freeing the mind from distractions to the task at hand. The basal ganglia, located deep within the subcortex or interior of the brain, dynamically modulates behavior based on feedback from the motor and affective systems and from our various sensory modalities (touch, vision, audition, gustation, and olfaction). These two brain structures are intimately connected and cutting their nerve fibers, as it turns out, is sometimes useful with patients with refractory OCD.

Indeed, malfunctioning basal ganglia lead to stereotyped movement patterns and the absence of novel behaviors. One patient I treated had obsessive thoughts that he was going to be infected with a sexually transmittable disease from casual sex (OCD) and another I observed in a clinic was a compulsive swearer (Gilles de Tourette syndrome). The first gentleman was a particularly interesting clinical case because he grew up in a home where his father was a compulsive gambler and continually stole money from his wife and two sons. As a result, the family was always broke, and family finances—like mortgage, electric, and gas payments—were often unmet. The twin brothers, who were identical twins, both suffered from OCD. The mother was the only "normal" individual in the family. It must have been quite difficult for her raising a family of compulsive gamblers and ideators. In both cases, the basal ganglia along with the orbitofrontal cortex, constrained the son's and the compulsive swearer's ability to switch mental set and each of them was mentally stuck in one mode or the other of responding to the world. The actions of the son led to certain thoughts and the thoughts of the compulsive swearer led to certain actions.

Clomipramine (Anafranil), a nonselective serotonin reuptake inhibitor (NSRI), can be an effective treatment for OCD. However, about 40% of patients with obsessive-compulsive disorder do not respond to either NSRI or selective serotonin reuptake inhibitor (SSRI) treatment (both of which raise serotonin levels in the brain). This may be so because some forms of obsessive-compulsive behavior are a result of excessively high levels of dopamine, not serotonin. High doses of dopaminergic drugs that increase production of dopamine in the brain (amphetamine, apomorphine, bromocriptine, and L-DOPA) appear to increase stereotypical movement and compulsive behaviors in humans in one of four dopamine pathways, the nigrostriatal pathway. On the other hand, another dopamine pathway, the mesocortical pathway, appears to be involved with some of the cognitive symptoms of OCD, such as obsessive thoughts. Genetic studies suggest that these kinds of obsessive-compulsive disorders are highly heritable. As a result, it is often difficult to treat OCD individuals with supportive psychotherapy alone, so treatment is sometimes augmented with the use of psychoactive drugs.

What about everyday risk-taking? Many forms of gambling and stock market speculation (investing in stock options and mortgage-backed securities), high-risk activities (bungee jumping, rock climbing, spelunking or cave exploration, and parachuting), and similar pursuits, probably arise from deep roots in human nature that are affected by culture, age (adolescence), and experience. They have many positive and some negative benefits and potentially lurk within all of us. To be sure, a sobering thought.

Psychosomatic Disorders

Historical periods affect what physical and mental symptoms are expressed in illnesses having psychological causes. Some medical professionals believe that the unconscious mind is the causative agent. Physical symptoms having a psychogenic origin are common and the processes that cause them in the mind-brain-body (MBB) are known as somatization. When there is actual physical disease, but the patient's responses are inappropriate or exaggerated, the processes that cause them are said to be somatogenic in origin. Four categories of physical and mental symptoms in which the mind-brain-body fashions diseases states can be identified with accompanying examples.

1. Sensory symptoms (tiredness, burning skin sensations)
2. Motor symptoms (paralysis)

3. ANS or autonomic nervous system symptoms (irritable bowel)
4. Psychogenic pain symptoms (headache caused by psychological factors)

The initial shaping of somatization began with the diagnosis of spinal irritation in the 1820s in which the patient's symptoms were culturally shaped with the physician acting as the agent of culture. For example, a condition known as reflex neurosis could cause any irritated organ to spread its influence to other parts of the body or brain. As women were considered the more passive gender at the time, they were considered more susceptible and were more often diagnosed by physicians with this disorder. Two new models of illness, however, arose after 1870. One stressed covert but actual disease in the central nervous system (CNS). Neurasthenia or tired nerves was a commonly diagnosed ailment.

> … the CNS becomes dephosphorized, or perhaps loses somewhat of its solid constituents; probably also undergoes slight, undetectable, morbid changes in its chemical structure and, as a consequence becomes more or less impoverished in the quantity and quality of its nervous force.[2]

Another stressed the psychological basis of somatization. In this instance, the mind creates actual physical symptoms, but the patient accepts them as signs of real physical disease. The experience of pain is a poignant example. "From the cultural pool, pain is selected from the symptom pool. But how people experience pain, how they describe it to others, and how and where they seek help is another thing."[3] In the United States, the most common form of somatization is chronic pain syndrome or fibromyalgia. Its origins have been variously attributed to chronic neurosis, infectious mononucleosis, Epstein-Barr virus, muscle weakness or neuromyasthenia, benign myalgic encephalomyelitis, and yeast infections.

The attribution of an illness involves two distinct phases. In the first phase, the patient appropriates a real physical disease as a *template* whose actual cause is difficult to substantiate. In the second phase, the patient broadcasts this template to others as an explanation for their set of symptoms. Broadcasting is abetted by solicitous physicians, pharmaceutical companies that stand to make money from selling expensive prescription drugs, patient support groups, and particularly, the mass media, which accrues income from increased advertisement revenue. The patient draws upon the pool of symptoms from this culture as models of illness to help them understand their conversations with their bodies. Social isolation and loneliness increase somatization and they are both correlated with poor health, actual physical disease, frequent visits to physicians, and persistent somatic complaints.

Sigmund Freud, the Viennese neurologist and psychoanalyst, believed that somatic complaints, usually expressed as some form of pain, were used by the mind-brain-body (MBB) to psychologically safeguard the individual from undesirable and socially unacceptable feelings. Thus, under Freud's diagnostic rubric they performed a *neurotic function* by keeping negative and objectionable feelings unconscious and inaccessible to the conscious mind. For instance, an irrational belief that one is a worthless person might generate perfectionist tendencies that lead to feelings of unmitigated and unconscious rage. The unconscious feelings themselves are suppressed because they are negative and socially undesirable. However, feelings just don't exist in some subjective space in the brain but are part and parcel of the physiological and bodily systems that give rise to them. And when these systems are under physical and psychosocial stress, they can physically alter bodily tissues leading to psychosomatic symptoms. On the other hand, when pain pathways to the brain are stimulated by real pain but there is no actual physical damage to the body, the symptoms are treated as hysterical, not psychosomatic. Dr. John Sarno of New York University, a noted specialist in this area, has identified a psychosomatic triad that underlies physical symptoms of the body caused by psychosocial stress. This triad includes narcissism or preoccupation with the self, feelings of inferiority in regard to others, and a tendency to be dependent on other people for emotional and social support.

When this system is overworked, the immune system tends to overreact contributing to a large number of immune-related problems including diseases of the skin such as acne, psoriasis, vitiligo, eczema, and hives as well as well-documented autoimmune disorders in which the immune system attacks one's own body found in some forms of multiple sclerosis and psoriasis, among others. On the other hand, *inflammatory memory* or memory of previous insults to the body, doesn't just reside in the immune system but also dwells in tissue stem cells. The ability of stem cells to regenerate tissues and potentially remember their past stressors accentuates the role of the mind-brain-body (MBB) working as an indissoluble unit.

Hypertension or high blood pressure is sometimes linked to repressed emotion. Such individuals or *repressors* as they are called, however, are unaware of their underlying psychological issues and are less likely to seek out psychotherapy or other forms of psychological or psychiatric treatment. And, while stress reduction approaches are useful in dealing with anger or anxiety, combination therapies with alpha and beta blockers are more useful in reducing the core medical problem of high blood pressure. Alpha and beta blockers actually reduce the effects of stress hormones such as cortisol and epinephrine on the body.

Tension Myositis Syndrome (TMS) is another psychological disorder with physical manifestations and it is often associated with chronic back pain, among other symptoms. Fifty percent of Americans complain of lower back problems along with accompanying pain. When very positive developments occur in people's lives, such as significant career changes—changing jobs, getting promoted or noteworthy alterations in one's lifestyle as in getting married—TMS symptoms typically reside. Yet negative emotions such as anger, anxiety, and depression as well as psychosocial stress can create painful conditions in the body. These bodily changes may redound to one's mental state potentially creating serious psychiatric conditions or unpleasant psychological issues in themselves. Recent research suggests that there are multisynaptic circuits that bond cognition, affect, and movement with associated coupling of the adrenal medulla and related autonomic and endocrine systems and they all make up as critical part of the stress response. The primary motor cortex (MC) is also involved as well as the axial musculature of the trunk and head. This suggests, however, that the core musculature in communion with the motor cortex and physical exercise, play a significant role in tamping down the stress response. Indeed, by experimentally following transneural transport of the rabies virus, two broad networks that impact on the adrenal medulla have been pinpointed. One network includes the motor areas of the frontal cortex and parts of the sensorimotor cortex. This network enlists the supplementary motor and cingulate motor areas for skeletal control of response selection, motor preparation, and the execution of movement. The other network includes parts of the medial prefrontal cortex and regions of the anterior cingulate cortex that are linked to *affective or hot cognition*. This occurs when someone's thoughts are strongly influenced by emotional factors, which would most likely be the case if you're affected by strong negative emotions not to mention psychological and physical stress. The opposite, of course, is *cold cognition*, which is a lack of emotional involvement such as calculating—you got it—someone else's taxes.

An Unconscious Mind?

We might digress a bit and ask ourselves whether there even is such a thing as the unconscious mind in the Freudian sense? Or, might it be conceptual shorthand for a phenomenon that in reality suggests a lack of scientific knowledge about how the mind stores information such as thoughts, feelings, ideas, events, facts, and so on? Some of these memory phenomena have a strong scientific basis. We know that memory for experiences, for instance, are stored in the brain in two fundamental ways: Explicitly and implicitly. Explicit memories include memory for events and associations we have about these events as well as memory for facts or the storage of formal, organized knowledge. The former might include remembrances of the day we got married or a running autobiography of our own life. The latter might comprise a collection of facts about where we were born, our age, gender, and the schools that we attended. Implicit memory, however, is a kind of memory in which there is storage of emotional and motor memories including motor skills, learned habits as well as involuntary behavioral responses or reflexes. Obviously, they lack the "pointillistic" quality of explicit facts and events.

Let's take a moment and look at this in more depth. There are three central processes of memory: Encoding, storage, and retrieval. Emotional memories are typically encoded with an emotional valence (how positive or negative the emotional episode was) as well as the context in which the memory was experienced. This context might include who we interacted with (our spouse, our next door neighbor's barking dog), under what conditions it occurred (over dinner, after arriving home from work, at a party) along with other incidental information such as the time of day, whether we were inside or outside, the color of the coat of the person to whom we were speaking, and so on. Emotional memories are initially processed in an archaic area of the subcortex, the amygdala. Memories that are particularly painful are then stored in the adjacent temporal lobes and the brain has a strong evolutionary incentive to do so because we are then reminded of the unpleasant consequences of a particular course of action under similar eliciting conditions in future contexts.

Our sense of repression of painful memories may actually be a retrieval problem as memories are not so much repressed as blocked from ready access along the original storage route often as the result of intervening experiences that distort or bias the mnemonic process. Sigmund Freud in his day called this poorly understood process, a *screen memory*. He understood it to mean a distorted memory of our childhood past that served to screen out traumatic events so that our conscious mind would be spared any psychic pain. But we know from research on memory by Professor Daniel Schachter at Harvard and others that prior beliefs and preexisting knowledge can influence our memories of the past. Our current emotional state and mood can distort our recollections of past events, too. If you are in a bad mood at the moment, your recollections of a past love might not appear as sanguine to your mind's eye as the actual relationship was 10 years ago as well as, unsurprisingly, the converse. Similarly, your current views on global warming might significantly influence your recollections of your past opinion on the matter, let's say 20 years ago, when most people didn't consider it as significant. There is a biological and psychological tendency in the mind-brain-body to bring into line your views twenty years ago, in this case, with your present beliefs, a phenomenon called *retrospective bias*. Nothing really here about repression or suppression of memories but rather a cultural distortion of your beliefs about the past based on fundamental properties of the brain, body, and central nervous system. The concept of repression, thus, is an antediluvian term of 19th century origin for brain and bodily processes that are only beginning to be understood and in a completely new way.

One way we might get a different perspective on the mind-brain-body (MBB) is to examine the role of communication in animals and the different conduits for communication that have evolved.

Notes

1. When I was in college, I took a one-year practicum at the Douglass Developmental Disabilities Center at Douglass College of Rutgers University, originally under the direction of Dr. Sandra Harris. I immediately related to autistic children who I found unbelievably fascinating. Ever since I have studied and done research on autism and worked clinically with autistic children and adults as a neuropsychologist.
2. The quote, "… the CNS becomes dephosphorized …" is from E. Shorter, *From Paralysis to Fatigue: A History of Psychosomatic Illness in the Modern Era*, 1992.
3. The quote, "From the cultural pool, pain is selected …" is from E. Shorter, *From Paralysis to Fatigue: A History of Psychosomatic Illness in the Modern Era, 1992.*

Bibliography

Alm, P. A. (2004). Stuttering and the basal ganglia circuits: A critical review of possible relations. *Journal of Communication Disorders*, *37*(4), 325–369.

Courchesne, E., Townsend, J., Akshoomoff, N. A., Saitoh, O., Yeung-Courchesne, R., Lincoln, A. J., … Lau, L. (1994). Impairment in shifting attention in autistic and cerebellar patients. *Behavioral Neuroscience*, *108*(5), 848–865.

Damasio, A. R. (2003). *Looking for Spinoza: Joy, sorrow, and the feeling brain.* New York, NY: Harcourt.

Dum, R. P., Levinthal, D. J., & Strick, P. L. (2016). Motor, cognitive, and affective areas of the cerebral cortex influence the adrenal medulla. *Proceedings of the National Academy of Sciences of the USA, 113*(35), 9922–9927.

Edelman, G. M., & Tononi, G. (2000). *A universe of consciousness: How matter becomes imagination*. New York, NY: Basic Books.

Gardner, H. (1993). *Frames of mind: The theory of multiple intelligences*. (Rev. ed.). New York, NY: Basic Books.

Freud, S. (1940). *An outline of psychoanalysis*. London: Penguin.

Handler, L. (1998). *Twitch & shout: A touretter's tale*. Minneapolis, MN: University of Minnesota Press.

Hazlett, H. C., Gu, H., Munsell, B. C., Kim, S. H., Styner, M., Wolff, J. J., … Zwaigenbaum, L. (2017). Early brain development in infants at high risk for autism spectrum disorder. *Nature, 542*(7641), 348–351.

Naik, S. (2018). The healing power of painful memories: Epidermal stem cells "remember" inflammation, accelerating subsequent wound repair. *Science*, 359(6380), 1113.

Sarno, J. E. (1998). *The mindbody prescription: Healing the body, healing the pain*. New York, NY: Grand Central Publishing.

Sarno, J. E. (2006). *The divided mind: The epidemic of mindbody disorders*. New York, NY: HarperCollins.

Schachter, D. (1999). The seven sins of memory: Insights from psychology and cognitive neuroscience. *American Psychologist*, 54(3), 182–203.

Seitz, J. A. (1996). *Social deficits in autistic children: The role of emotion and empathy*. Unpublished manuscript, Department of Political Science & Psychology, York College, City University of New York (CUNY).

Seitz, J. A. (2000). *The embodied self*. Paper presented at the 30th Annual Symposium of the Jean Piaget Society, Montreal, Canada.

Shorter, E. (1992). *From paralysis to fatigue: A history of psychosomatic illness in the modern era*. New York, NY: The Free Press.

· 6 ·

COMMUNICATION

In this chapter, I examine prelinguistic communication in human infants and various animal species; gesture, posture, facial expression, and vocal prosody; and Braille.

Communication in Sign

The use of sign language as a medium of communication illuminates the role of movement in language use. American Sign Language (ASL) depicts complex linguistic structure by encoding it in spatial contrasts by way of the hands and the body. That is, one might say that the body is both a vehicle for, but is also constitutive of, thought. ASL uses the spatial relationships of the hands and body to depict syntactic information such as verb objects and nouns by manipulating loci of the hands and body and relations among these loci in the immediate plane of signing space. Indeed, deaf signers with damage to the left hemisphere show the various language aphasias just like in spoken language.

Right hemisphere damage, however, results in distortions of space and spatial perspective and neglect of the left side of space known as *spatial neglect*. Objects in the left visual field in both eyes are processed by the right hemisphere and objects in the right visual field in both eyes are managed by the left hemisphere. As a result, patients with right hemispheric damage are unable to recognize objects in their left visual field such as in viewing photographs of office layouts. However, competence in ASL—that is, signing with the hands—is not affected suggesting similar brain organization for both sign and spoken languages. Unfortunately, the suppression of sign language use among the deaf has historically resulted in a significant deterioration in the intellectual achievement of deaf children and, developmentally, in delays in cognitive and social development.

Like spoken language, sign language is capable of expressing nonliteral meaning by the overlapping, blending, and substitution of signs as in deaf theatre or sign poetry. Examples of nonliteral meaning in spoken language include linguistic tropes like metaphor ("The smell of her perfume was like bright sunshine"), analogy ("The structure of the atom is analogous to the structure of our solar system"), and irony ("It was so cold it felt hot"). Using form and design in space that is superimposed on signs and signing, poetic structure, such as alliteration and assonance, is possible. In speech, alliteration occurs when a succession of words has the same sound in the initial consonant such as "Gary's giraffe gobbled gooseberries greedily." You can alliterate in sign language using a series of initial gestures that are identical. In assonance, the sound of a vowel or diphthong (two vowels in a single syllable that merge into one another as "oi" in the word "coin") is repeated in stressed syllables such as "penitence, reticence" and you can also duplicate this phenomenon in sign with identical stressed gestures. Undoubtedly, the use of ironic and metaphoric modes of communication occurs frequently in deaf communities. But figurative sign language is also produced in deaf children through pantomime, sign modification, ritualized movement, and by adding iconic, visuospatial detail to their signed expressions.

Nonverbal Communication

We communicate with our bodies through gesture and posture or what's known as *kinesics*, but we also communicate about the spatial context in which we interact with others. That is, we communicate with others through movement of our limbs as well as by way of whole-body movements, but we also commune with others by the way we interact with them within the immediate space around our bodies. This is called *proxemics* and concerns the use of space by us and other animals and how it is affected by our behavior; the density of people, objects, and things around us; and the kind of social interactions we are engaged. We also convey meaning to others through symbolic action such as use of the American flag to criticize the government or support it. We communicate with our bodies through touch; paralinguistic cues such as vocal prosody—tone, stress, rhythm, pitch, intonation, and volume—as well as through *chronemics* or time perception. Use of time might include being punctual with a friend or colleague, a willingness to listen or wait, conveyance of social status, and so on. There are many ways that humans and other animals communicate with each other and vocal languages are just one of them.

Indeed, nonverbal behavior is central to expression and communication. Infants and young children learn to communicate with gestures before they learn to speak, and this mode of communication continues into adulthood where a large body of kinesic behaviors augment or replace language such as illustrators, regulators, affect displays, deictics, metaphoric gestures, emblems, and a huge class of procedural knowledge and skills. Of the latter, choreography and dance, sports, and craftsmanship are a few examples of procedural skills. Of the former, *illustrators* are gestures that serve to demonstrate what is being said in a verbal statement such as using hand gestures to denote the size and shape of an object. *Regulators* are nonverbal signs that modulate conversational interaction such as a nod of the head. *Deictics* are, of course, pointing gestures, and *emblems* directly refer to their verbal equivalents such as a hitchhiker's raised thumb.

With regard to scanning the printed page with one's fingers as in the blind, *Braille* is essentially the reading of a tactile code in which the number and spatial forms of the raised dots is critical. In human cultures, facial expression, gesture and posture, gaze, spatial behavior among conspecifics, touch, bodily movements, vocalization, smell, and appearance are essential and basic to communication and in many animal "cultures," too. Body therapies such as the Feldenkrais technique, Hatha yoga, and massage therapy, which I discuss in Chapter 10, are also examples of nonverbal abilities.

Some observers have argued for the gestural origins of language and the fact that both speech and hand control originate from the same neural systems. As we saw earlier there is, in fact, a close synchrony between gesture and speech. Michael Corballis, a cognitive scientist, suggests that the restrictions required of control of the breadth and the rhythmic character of articulation would have entrained gestures to the speech articulatory apparatus as the two evolved, more or less, together. Concepts originating from gesture that were used to depict actions and objects—such as moving the fingers across space to depict a car or enclosing the hands to depict a cup—were eventually reified in symbols such as words, a position that the Swiss psychologist, Jean Piaget, took at mid-century. That is, concepts essentially evolved from movement, action, and activity. Or, in Piaget's language, movement was internalized as thought and the formation of mental concepts. Corballis argues that the basic structure of language is spatial in nature and represented in the inferior parietal lobule (IPL), which in human development first directs grasping followed by speech. But, the crux of his argument is that speech may have evolved much earlier than previously thought and is found not only in early modern humans but in some common ancestor of Denisovans and Neandertals, both archaic hominins.

Other observers have noted that speech and gesture appear to have evolved, more or less, together. Mirror neurons, as we discussed earlier, essentially map the observation of actions with the execution of actions. But they had three consequences in evolutionary terms. They enabled complex sequential motor actions, the recognition of both manual gestures of the body and affective displays of the face, and the ability to pantomime. The latter appeared to have evolved initially from visual communication with others by way of facial expressions and from there to the use of manual gestures of the body and finally to the expression of gestures of the voice. That is, we initially mimed people's expressions, then we pantomimed their bodily gestures and postures, and finally their vocal characteristics. This is particularly useful if you are a very talented actor like Sairoise Ronan (pronounced "seer-sha" in Ireland meaning "freedom" in Gaelic, but "sur-shuh" in American English), the 24-year-old Irish-American actress, who possesses an uncanny ability to mimic the facial expression, gesture and posture, and voice of others.

These gestures of the voice are referred to as *phonetic gestures* and purportedly evolved from face-mouth mirror neurons in area F5 and found in a number of primates, not just hominins. With the recognition and generation of sounds meaningfully coupled to communicative gestures the full panoply of speech functions reached its present zenith. Nonetheless, mirror neurons are not, in principle, able to denote the social intentions of an individual and cannot account for the complex properties of human languages, particularly the meanings of words and sentences.

Hominids

Let's step back for a moment and look at the *hominid* line (gorillas, chimps, orangutans, and humans). Ape gesture is tied to intentional communication and serves various communicative functions whereas vocalizations in apes are more reflexive and lack communicative intent. Thus, one could draw the conclusion that communication through gesture is a much older phenomenon that dates back to Great Ape culture and early hominins who evolved 1.5–2.0 million years ago (mya). At this time, an early hominin, *Homo erectus*, would have required some form of enhanced communication because it developed a complex tool industry known as Acheulian culture involving the manufacture of bifacial axes, among other things (Figure 6.1). These early hominins shaped stone by flaking it and used not just stone flakers but wood, antler, and bone. They worked the stone symmetrically indicating a burgeoning understanding of early technology. Gesture would have served this culture well, particularly in teaching and learning. They also made use of fire including the use of heaths for cooking and presumably made use of the digits for indexical pointing or the finger-pointing gesture.

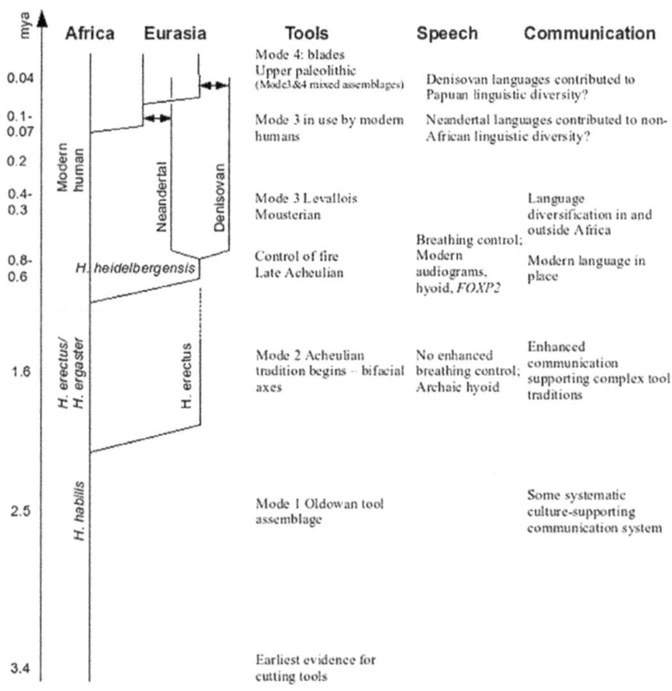

Figure 6.1. Modern Human Evolutionary Ascendance.
Credit: Dediu D and Levinson SC (2013) On the antiquity of language: the reinterpretation of Neandertal linguistic capacities and its consequences. *Frontiers in Psychology*, 4 (397). doi: 10.3389/fpsyg.2013.003971. Copyright notice: This is an open-access article distributed under the terms of the Creative Commons Attribution License, which permits use, distribution and reproduction in other forums.

It has been proposed that *H. erectus* possessed a processing system for creating sentences, that is, some form of syntactic carpentry, although not necessarily the use of a universal grammar (UG) that putatively evolved later. They also made use of *primitive metaphor*, which I have argued is an evolved biologically-based system such as relating a smell to a color or a texture to a sound. Then, there is the emergence of mutations on the FOX2P gene that were concurrent with the emergence of early *H. erectus* and were probably fully up-and-running something like 260,000 years ago (260 kya) or as early as 500,000 years ago (500 kya). It is believed that mutations of the FOX2P gene supported precision timing of the vocal apparatus as well as brought on the ability to orchestrate acoustic elements of speech in correct temporal order. Indeed, the first symbolic words—sounds that referred to a person, object or thing—may have arisen even earlier in *Ardipithecus ramidus* about 4.4 million years ago. Archeological studies suggest that *A. ramidus* had a lengthier socialization period in childhood than other primates and a more vocally-mediated sociality. That is, they evinced more elaborate vocalic abilities than other primates at the time such as bonobos or chimpanzees.

But more complex language did not arise directly from use of the hands as Professor Corballis and others have proposed, but from adaptation of the motor articulatory apparatus over evolutionary time for speech production. Recent evidence suggests that articulate speech first emerged in a common ancestor of Neandertals and Denisovans, both archaic hominins. This common ancestor is mostly likely *Homo heidelbergensis* who evolved from *H. erectus* and *H. ergaster* and who appeared on the scene approximately 600,000–800,000 years ago suggesting that speech may have been in place as early as half-a-million years ago. It was also embodied. It made use of the mouth, face, and hand as an integral unit to communicate with others as do current sign languages, even though they are more hand than mouth, so to speak, as in speechreading or lipreading. *Speechreading* makes use of visual clues from movements of the lips, bodily gestures, and facial expressions to communicate with others. *H. heidelbergensis* took advantage of fire and possessed a late Acheulian culture of toolmaking, particularly the Levallois technique, which is a unique form of stone knapping or fashioning tools by use of hard rocks to chip away and shape them. This particular archaic hominin appeared to have had the essential genetic and biological components for speech such as the basic framework of the speech articulatory mechanism and the ability to control the breathing apparatus in order to intone words in real time.

It is true that a number of primates, mammals (whales, dolphins), and songbirds have considerable vocal plasticity so that they can acquire novel kinds of sounds and even modify their communications when there are changes to their social environment. For instance, the vocal tracts of baboons as well as macaques—small Old-World monkeys living in intricate social structures—can generate human-like vowels across the full-range of formants. Formants are components of acoustic signals that possess distinctive frequencies. Vocal and gestural turn-taking is also fairly common in bonobos and chimpanzees and either one or the other also occurs in other primates. It is a key component of human speech. Vocal turn-taking is seen in marmoset infant monkeys (small, New World monkeys), squirrel monkeys, and lemurs, among others. Indeed, in marmosets, parents shape turn-taking starting in infancy. Similarly, many primates deftly engage in referential signaling or the use of vocal or manual gestures to refer to external objects or happenings in their environments. They also share resources with others within their social groups. Tamarins and capuchins share resources, both small New World monkeys. Capuchins, by the way, are the organ grinder monkeys often seen in movie flics and on TV. Overall, this suggests a number of interrelated abilities including communication and referential signaling that are combined in the sharing episodes. Then too, there is the suggestion that archaic early humans used language, at least initially, to synchronize and harmonize social interactions in large groups of individuals. So, language evolved for a number of reasons: To convey our thoughts, tools for thinking, to communicate with others our feelings and desires, and bond larger social groups.

Canines, Cephalopods, and Parrots

If we take a look at canines, *Canis familiaris*, the ability to evaluate and process both intonation and word meaning encountered in interactions with the human species evolved in the complete absence of the possession of human-like languages. This implies a number of things. The processing of words or *lexemes* (thus, lexical processing) in the left hemisphere may have evolved in a common ancestor of canines who began to link sound to meaning. In modern canines, there is a left hemisphere bias for understanding words and a right hemisphere bias for discriminating intonation of words as opposed to other sounds. Non-auditory regions of the canine brain then evaluate this linkage for meaning. So, human praise is sized up within the mesolimbic dopamine system or reward pathway because canines depend on the reward value of human utterances and that information is processed in the ventral striatum, substantia nigra, and ventral tegmentum where dopamine cell bodies lie. Similarly, affective aspects of the voice and other channels of affective communication, such as gesture and posture of the body, are analyzed in the right middle ectosylvian gyrus in a similar location as the auditory cortex of the temporal lobe in humans. What's interesting in all of this is that early on, in the mutual enculturation of canines and humans, domestication may have had a significant effect on the actual processing of words and affect (emotions) in the canine mind-brain-body (MBB). In fact, some breeds of canines have extremely good *recognition memory* for words and can use up to a thousand of them as discriminative stimuli for previously encountered individuals, objects or events. A border collie I mentioned earlier can reputedly retrieve named objects from a large array of familiar objects and will even retrieve an unknown object when given a previously novel name not heard before.

In cephalopods, who evolved 530 million years ago such as the octopus, the cuttlefish, and the squid, the skin contains chromatophores and iridophores that produce pigments including red, yellow, and black/brown. The skin also contains reflecting cells called leucophores. The octopus putatively communicates with fellow octopuses and other sea creatures such as predators (sharks) by creating deimatic displays or dramatic patterns on the surface of the skin to, in all likelihood, startle predators as it scoots away. In squids, there are up to thirty ritualized displays of changing patterns and sequences. The interpretation of these complex patterns on the surface of the skin of cephalopods, in general, is that they constitute a *visual language* that may even possess an inchoate grammar, that is, sequenced patterns for communicating with others. Cephalopods have large nervous systems of half a billion or more neurons and their brains are capable of extensive learning and memory as well as complex behavior. However, in order to survive in the wild, particularly the octopus who lacks a shell or tough outer surface, it appears that the mind-brain-body (MBB) diversifies the proteins expressed by cephalopods by way of RNA editing and extensive recoding of messenger RNA. The gene family, Protocadherin, are expressed in both the mammalian brain and in the nervous systems of Cephalopods, but in the latter, it is greatly expanded and believed to play a role in diversifying neural circuitry as well as increasing the complexity of neural connections. RNA editing has several advantages as it can be quickly turned off and on unlike changes to the underlying DNA.

African grey parrots (*Psittacus erithacus*) also possess some engaging communicative abilities. They can put together sentences constructed out of diverse labels for words that they have learned from their trainers, which suggests that they may interpret the speech of humans as made up of individual units. They can generalize labels for objects to unfamiliar situations and then recombine these labels to name new objects, things, and people. It has been noted by Irene Pepperberg, the noted animal cognition researcher, that parrots can form a new word such as "banerry" by combining the words "banana" and "cherry." But it also looks like they possess the capacity for recursion. Recursion is informally defined as "a process a procedure goes through when one of the steps of the procedure involves invoking the procedure itself." When asked, "What color is the item that is circular and rawhide?" parrots are able to generate a *logical conjunction* to connect features of objects (circular and rawhide) after dividing the question into its parts (color, shape, texture) and repeatedly using each part to formulate their answer. In ordinary or natural language processing, a sentence can be structured so that whatever follows the verb is just another sentence. In the sentence, "Barbara believes that rabbits are smart," the first noun and verb form a sentence with an object, "Barbara believes ______," that is followed by another sentence, "Rabbits are smart," that is an object of the first. So, the second sentence is embedded within the first and is formed recursively. This has been considered to be a fundamental property of language, mathematics, music, and other symbol and action systems. That's because it is an essential property of movement, as I argued in the first chapter.

What goes on in parrot speech when they are left alone without someone or a conspecific to communicate with? So-called crib play, as it is called in human infants, paints a rare picture of the unattended child or parrot, in this latter case, learning speech. This kind of unattended vocal play is common in wild birds and young parrots will repeat complete dialogues when left alone. Their *crib monologues or sound play* involves the imitation, variation, and modification of vocal sounds in the parrot's past social environment. Like canines, they engage in fast mapping or *referential mapping* of words ("blue") to referents (blue color) when learning new names for things. This is unlike *mimicry* when parrots simply mimic sounds in the environment, which lack any reference, functionality or interactive social environment. They also appear to retain an elaborate acoustic map between their speech and the speech of their human trainers, at least in laboratory studies. Importantly, while birds diverged from mammals about 280 million years ago, the avian brain has remarkable similarities to the hominid brain with a similar underlying cortical architecture.

Forms of communication are widespread in the animal kingdom although not the complex language found in human cultures with layers of syntax and meaning. But it exists, predates human speech, and while the specifics are far from fully worked out, it's where the origins of human language lie.

Next, we will turn to the embodied nature of art.

Bibliography

Andics, A., Gabor, A., Gasci, M., Farago, T., Szabo, D., Miklosi, A. (2016). Neural mechanisms for lexical processing in dogs. *Science*, *353*(6303), 1030–1032.

Arden, R., Bensky, M. K., & Adams, M. J. (2016). A review of cognitive abilities in dogs, 1911 through 2016: More individual differences, please! *Current Directions in Psychological Science*, *25*(5), 307–312.

Barton, S. S. (1979). Development of gesture. In N. R. Smith & M. B. Franklin (Eds.), *Symbolic functioning in childhood* (pp. 139–151). Hillsdale, NJ: Lawrence Erlbaum.

Bates, E., Thal, D., Whitesell, K., Fenson, L., & Oakes, L. (1989). Integrating language and gesture in infancy. *Developmental Psychology*, *25*(6), 1004–1019.

Bickerton, D. (2014). *More than nature needs: Language, mind, and evolution*. Cambridge, MA: Harvard University Press.

Corballis, M. (2011). *The recursive mind: The origins of human language, thought, and civilization*. Princeton, NJ: Princeton University Press.

Corballis, M. (2017). *The truth about language: What it is and where it came from*. Chicago, IL: The University of Chicago Press.

Ekman, P., & Friesen, W. V. (1969). The repertoire of nonverbal behavior: Categories, origins, usage, and coding. *Semiotica*, *1*(1), 49–98.

Fox, K. C. R., Muthukrishna, M., & Shultz, S. (2017). The social and cultural roots of whale and dolphin brains. *Nature Ecology and Evolution*, *1*, 1699–1705.

Fugazza, C., Pogany, A., & Miklosi, A. (2016, November 23). Recall of other's actions after incidental encoding reveals episodic-like memory in dogs. *Current Biology*, *26*(23), 3209–3213.

Godfrey-Smith, P. (2016). *Other minds: The octopus, the sea, and the deep origins of consciousness*. New York, NY: Farrar, Straus & Giroux.

Goldin-Meadow, S. (1997). When gestures and words speak differently. *Current Directions in Psychological Science*, 6(5), 138–143.

Hardman, M. L., Egan, M. W., & Drew, C. J. (2017). *Human exceptionality: School, community, and family*. (12th ed.). Boston, MA: Cengage Learning.

Heller, M. A., & Ballesteros, S. (2006). *Touch and blindness: Psychology and neuroscience*. Mahwah, NJ: Lawrence Erlbaum.

Hochner, B., Shomrat, T., & Fiorito, G. (2006). The octopus: A model for a comparative analysis of the evolution of learning and memory mechanisms. *Biological Bulletin, 210*(3), 308–317.

Iverson, J. M., & Goldin-Meadow, S. (1998). Why people gesture when they speak. *Nature, 396*(6708), 228.

Jones, S. (2005). The role of mirror neurons in imitation. In S. Hurley & N. Chater (Eds.), *Perspectives on imitation: From neuroscience to social science*. Vol. 1. Mechanisms of imitation and imitation in animals (pp. 205–209). Cambridge, MA: MIT Press.

Kaminski, J., Call, J., & Fisher, J. (2004). Word learning in the domestic dog: Evidence for "fast mapping." *Science, 304*(5677), 1682–3.

Klima, E. S., & Bellugi, U. (1980). Wit and poetry in American sign language. In W. C. Stokoe (Ed.), *Sign and culture* (pp. 105–132). Silver Spring, MD: Linstok Press.

Liscovitch, N., Alon, S., Porath, H. T., Elstein, B., Unger, R., Ziv, T., … Eisenberg, E. (2017). Trade-off between transcriptome plasticity and genome evolution in cephalopods. *Cell, 169*(2), 191–202.

Marschark, M., Everhart, V. S., Martin, J., & West, S. A. (1987). Identifying linguistic creativity in deaf and hearing children. *Metaphor and Symbolic Activity, 2*(4), 281–306.

McNeill, D. (1992). *Hand and mind: What gestures reveal about thought*. Chicago, IL: The University of Chicago Press.

Nicoladis, E., Mayberry, R. I., & Genesee, F. (1999). Gesture and early bilingual development. *Developmental Psychology, 35*(2), 514–526.

Pepperberg, I. M. (1999). *The Alex studies: Cognitive and communicative abilities of grey parrots*. Cambridge, MA: Harvard University Press.

Piaget, J. (1952). *The origins of intelligence in children*. (2nd ed., M. Cook, Trans.). New York, NY: International Universities Press.

Pinker, S., & Jackendoff, R. (2005). The faculty of language: What's special about it? *Cognition, 95*(2), 201–236.

Sacks, O. (1990). *Seeing voices: A journey into the world of the deaf*. New York, NY: HarperCollins.

Snowdon, C. T. (2017). Learning from monkey "talk": Nonhuman primates are capable of several key elements of language. *Science, 353*(6330), 1120–1122.

Sternberg, M. L. A. (1999). *The American Sign language Dictionary* [CD-ROM]. Seattle: Multimedia 2000.

· 7 ·

ART

Is There an Aesthetic *Sense*?

Pablo Picasso once said that "we have discovered nothing new in art in 17,000 years."[1] If Picasso was right and nothing really new has been added to representational art since the Magdalenian period in early human history eighteen thousand years ago, then the practice of aesthetics and the evolution of a human aesthetic impulse—opposed to the formal study of art and beauty—has a very long history. Bas-relief sculpture, perspective drawing, molded clay, sculpture in the round, engraving, shading and false relief, oil- and water-based paints, fired ceramic figures, charcoal drawing, stamping, stenciling, and brush painting all existed by this period. Might there have been a special aesthetic *sense* that evolved in humans?

What would an aesthetic *sense* comprise? What might differentiate an aesthetic *sense* from the simple cultivation of discrimination or taste, as David Hume, the 18th century British philosopher, proffered. Is the aesthetic *sense* tightly coupled to the individual sensory modalities on the presumption that taste and discrimination are, at the very least, based on the cultivation of the senses? If so, what might the aesthetic *sense* consist? For instance, in dance, might it be closely coupled to proprioception and the vestibular senses? In the culinary arts or winemaking, does it represent the fine honing of gustation and olfaction given the putative importance of both senses in the appreciation of food and drink? In music, is it principally a faculty of audition? In painting, sculpture, architecture, fashion, and film does it primarily rest on the eye and associated visual apparatus? In the theatrical arts, does acting and directing involve some combination of proprioception, audition, and the haptic sense?

Alternatively, is such an aesthetic *sense* a proxy for institutional practice that educates the senses? The disciplinary or occupational domain transmits and shapes artistic abilities, skills, and products. Such domains include the choric and culinary arts, painting, sculpture, the crafting of buildings, fashion, filmmaking, music-making, and acting. One might think of a domain as a "stable cultural realm that preserves and transmits new ideas or forms."[2] For example, one can learn the complex skills of a pastry chef employed at an upscale restaurant by immersing oneself in the day-to-day activities of the pastry making: The biology of and chemistry of foodstuffs, the technology of baking, the art of glazing, and so on.

Nevertheless, it is the collection of social institutions or, rather, the social and cultural organization of a domain, that legitimates the creative products of an individual artist or creator. These social institutions encompass collectors and philanthropists, critics and curators, museum directors and gallery owners, scientific review panels, professional and trade associations, and industry and academic leaders. Thus, the field can be conceptualized as a panoply of social and political institutions that "select from variations produced by individuals those worth preserving."[3] Pablo Picasso's first significant works of art as a young artist were initially recognized and promoted by a small group of patrons in the *Artworld*. In advanced capitalist cultures the individual is situated within a matrix of market relations and market competition. Artistic creativity emerges within a larger social and institutional matrix in which visual and creative ideas are commodities and their value in the marketplace is both galvanized or suppressed by existing politico–social organizations and institutions.

I want to consider in this chapter what role the mind-brain-body (MBB) might play in the functioning of an aesthetic *sense*. Immanuel Kant, the 18c German philosopher, believed that the operation of the higher cognitive functions shape what the senses transduce. The notion of an aesthetic *sense* may thus be a misnomer ignoring the central importance of the operation of the higher cognitive functions on the information that the senses provide. We might ask what role does beauty play in an aesthetic sense? Is it a surrogate for some property of apperception—a synthesizing or constructive activity of the mind—that intervenes in the interpretation of a percept? If so, is there such a thing as an aesthetic universal that permeates all mind-brain-bodies? I explore these various questions, in turn, below.

Initially, Kant regarded the aesthetic *sense* an intuitive and pleasurable experience that guided human conduct. He maintained that an aesthetic experience was not just a sense pleasure but a property of subjective intuition or inference. He linked his central aesthetic concepts—the sublime and the beautiful—to the classic temperaments of melancholy, sanguine, choleric, and phlegmatic. The *beautiful* aroused joy in the observer such as viewing a bouquet of red roses. The *sublime* aroused awe and wonder such as viewing an evening sunset over the ocean. A *sanguine* individual, who is cheerful and optimistic, would be attracted to the beautiful but only the surface appearance of people and things. On the other hand, a *melancholic* person, who experiences dysphoria and sadness, would be attracted to the sublime such as the tragic plays of Shakespeare. In his later work, Kant argued that taste was not an intuitive but a logical judgment, uniting understanding, judgment, and reason and was founded on one that was aesthetic but undergirded by feelings of pleasure and desire. Thus, for Kant, an aesthetic response resided in the faculty of judgment and in the higher cognitive functions and did not originate directly in the senses.

Is There an Aesthetic *Sense* Underlying Dance?

Proprioception, the body's internal feedback with regard to the position of its limbs, is linked to special kinds of nerve cells in the brain called mirror neurons that may provide the basis of the aesthetic *sense* in dance. Aristotle, of course, argued that there were five external senses. In reality, however, there are also myriad internal senses. One of these, proprioception, provides feedback to the brain from sensory receptors in the muscles, joints, tendons, and skin that makes available to the brain information about the position of the limbs in space and time and the weight and position of objects with which the body interacts. Mirror neurons in the premotor cortex receive information from the somatic nervous system, principally information from the proprioceptive and vestibular senses, enabling humans to model external events, such as things seen or heard in self-initiated motor behaviors, as in copying the movements of others or mimicking auditory rhythms through rhythmic movements of one's own.

Nevertheless, such an aesthetic response is underrepresented by proprioception linked to mirror neurons in the brain. For one thing, we don't necessarily experience an aesthetic response in everyday, garden-variety motor activity and feedback, so proprioception and mirror neurons can't be doing all the work of any genuine aesthetic response. Proprioception and mirror neurons are necessary, but not sufficient conditions, for imitative behavior, namely, the conversion of visual input into motor output. The origin of imitative behavior—a central component of aesthetic movement—is not fully understood. Nevertheless, two cortical mechanisms for imitation have been recently proposed to account for imitative behavior. The *opercular region in the inferior frontal cortex* seemingly codes the observed action with respect to a movement's intended goal and the *right superior parietal cortex* appears to code the kinesthetic components of movement.

On the other hand, the vestibular system is immensely important to human movement and cannot be discounted. It provides information about the position and motion of the whole body in space principally through specialized motion cells in the inner ear. These mechanoreceptors as they are called, impart information to the brain and spinal cord about self-motion, head position, and the spatial orientation of the body relative to gravity, as well as contribute to the stabilization of posture, gaze, and position of the head in relation to the body as a whole. The vestibular-ocular reflex focuses the eyes on visual targets as the head moves, an important component of dance movement. Indeed, the integration of the vestibular and somatic systems as well as input from the visual system is central to movement of the whole body in space not just proprioception alone.

Nevertheless, following Kant's thesis that an aesthetic response is inexorably grounded in cognition and not an intuitive response based on fashionable appeals to the senses, the higher cognitive functions are no doubt central to, and indispensable for, the coordination and modulation of complex sequences of movements as well as their precise timing and spatial accuracy and comprise the core of aesthetic expression or appreciation. The senses, in terms of any individual *sense*, are thus accessories to an aesthetic response. Proprioception—really a gloss for mechanoreceptors in the muscle spindles, Golgi tendon organs, and joint and cutaneous receptors—provides feedback to the brain and spinal cord with regard to musculature force, tone, and length and comprise the so-called stretch or deep tendon reflexes.

Even in rare cases where an individual has been deprived of a *proprioceptive sense*, this does result in individual lacking an aesthetic response when viewing an artistic work. Nor does such an individual lack embodied concepts such as the notions of up/own or in/on. This is so because one is not deprived of an aesthetic *sense* as there are other intact sensory modalities, both internal and external, that provide sensory information and feedback. In the rare case of a patient deprived of proprioception from the neck down, the individual was able to draw as well as write poetry. His movements were consciously controlled by higher-order cognitive faculties with sensory feedback primarily from vision and audition. Then there is the case of Ian Waterman, a 49-year-old male deprived of a proprioceptive sense at the age of nineteen, due to an unusual disorder that resulted in the degeneration of afferent pathways from the body to the brain but spared efferent pathways controlled by the motor system to the body. He could see and visualize his body but couldn't sense its movement. He learned to move by learning to control his body's movements by use of his vision. He capitalized on another sense.

> Ian's enjoyment of any kinetic melody—in a child, a puppy, a ballerina—may have been heightened by his illness. Rather than making him shun the beauty of movement, his experiences seem to have given him a finer and deeper appreciation of it. He had to think about his movements far more intensely than any choreographer. His interest in such graceful animals as deer is hardly a coincidence.[4]

Similarly, when limbs are lost one is not deprived of an aesthetic response to dance movement. The brain generates phantom limb sensations when limbs are missing and does so by remapping cortical areas. An area in the brain comprising deep thalamic nuclei reorganizes itself to take over a former area that subserved the missing limb. The missing limbs, then, are really not missing at the level of the higher cortical centers.

Likewise, a little-known phenomenon called *corollary discharge or reafference* that we discussed earlier, presets the sensory system to anticipate motor acts. Higher cortical centers not directly connected to motor acts, the frontal lobes, monitor sensory feedback for a brief moment just before the motor cortices produce movement. Loss of frontal lobe function disrupts sensory feedback and radically affects movement production. Conversely, Gilles de Tourette Syndrome, which is characterized by uncontrollable facial and bodily tics and vocalizations as well as unremitting exploration of the environment, is often accompanied by heightened cognitive abilities as a way of handling difficult social situations as real-life "touretters" attest.

Like the proprioceptive system, the vestibular system, which receives feedback from mechanoreceptors in the inner ear, makes available to the brain and spinal cord information about the position of the head and body relative to outside influences on body position and stability. The actual planning and initiation of voluntary movement, however, takes place in the motor cortices, including the primary motor, premotor, and supplementary motor areas together with support from the basal ganglia and the cerebellum, subcortical structures.

In addition, there is mounting evidence that motor abilities are directly correlated with the volume of cortical tissue subserving those abilities in the brain. Increased use of a body part has been directly linked with greater representation of that body part in the cortex. In music, for example, left-hand finger digits in right-handed string players have a larger representation in the motor cortices than in non-musicians. When listening to well-practiced piano music, professional pianists exhibit involuntary motor activity and engage the primary motor cortex. Thus, in a theoretical vein, the philosopher, Alva Noë, contends that our perceptual experience is dependent upon the use of our sensorimotor knowledge and that we make use of our bodily skills to take advantage of the intrinsic relations between our bodies and our brains. In a similar fashion, the philosopher, Derek Melser avers that mental concepts are nothing more than actional concepts, that is, thinking is a learned voluntary action.

In the case of dance, an aesthetic response would not principally arise from some aesthetic *sense* or derive inexorably from some particular sensory modality but originate in mental events or higher cognitive functions. Previously, in Chapter 1 I described four cognitive abilities central to aesthetic movement: Motor logic and organization and kinesthetic memory and awareness. Motor logic and organization comprise the dancer's neuromuscular skill with regard to the assembling, articulation, and ordering of movement. This comprises the syntax of movement. Kinesthetic memory enables the dancer to think in terms of movement by mentally reconstructing muscular effort, movement, and position in space and time. Kinesthetic awareness, or the kinesthetic sense proper, an amalgam of vision, proprioception, and the vestibular sense, provides the dancer with online information about the extent, direction, and weight of movement. All four comprise the cognitive basis of an *aesthetic sense*. This *aesthetic sense* would include the ability to use one's body to convey an idea or emotion, create a spatial design, or fashion a more extended choreographic product in time and space. That is, there is no unique *sense* that encompasses individual art forms such as dance or music. Rather, the *aesthetic sense* is a property of the higher cognitive functions constituted in mutual communion with the body.

Keeping together in time, as William McNeill discusses in his book of the same name, executed through human group movement, had important historical consequences because, among other things, it facilitated human solidarity and social cohesion through such *enactive vehicles* as community dancing, marching, and singing; religious rituals; courtship; close-order military drill and battle; calisthenics; martial arts; sports; demonstrations; parades; choral singing; as well as some forms of factory labor. But it also enhanced and expanded the storage and retrieval of information, or what I'm going to refer to as *collective memory*, through group participation in large informational networks. Similarly, these group interactions augmented *collective intelligence* by emphasizing collaborative movement's cognitive function, which encouraged *participatory democracy*—participation in collective decision-making—and the facilitation of an individual's unique social identity.

Let's step back a moment and consider how an aesthetic impulse might arise in an individual dancer or choreographer. First, a dancer would match body parts of self and other and then generate movements mapped onto his or her own body configurations. Not surprisingly, human infants readily map their mental experiences to bodily acts as well as note equivalences between perceived and executed actions. Infants also assume that others have the same mental experiences and map mental properties onto behavioral states of others. While the *right inferior parietal lobe* picks out and detects the actions of others and maps them onto bodily aspects of self, the *left inferior parietal lobe* computes the obligatory sensorimotor relationships for imitation. Most dances—whether ballet or modern dance—are choreographed directly from observing others or by observing digitized videos of dances. Martha Graham's great modern dance works of the 1930s, *Lamentation* and *Primitive Mysteries*, were reconstituted from watching the steps on film as recounted to me by dancers in the company. I have directly observed modern and ballet dancers learn completely new choreographic works by watching it on a laptop computer screen positioned obliquely to their body, so they were literally on stage with the other dancers. Frankly, when a modern dancer trained in ballet first demonstrated it to me, I couldn't believe my eyes. They pickup movement seemingly instantaneously because they are already deeply embedded in the grammar of movement of a particular style of dance; Graham technique or French or Russian ballet method, for example. Even when dances are recreated from dance notation, the actual movements are worked out among dancers in the studio, as I observed and participated in, at the Laban Institute in New York City. *Labanotation* (1928), as it is called, was one of the first dance notations and created by Rudolf Laban who started the Institute named after him.

Second, these equivalence relations among mapped body parts create kinetic signatures of self and other that a dancer capitalizes on in movement. *Kinetic signatures* are correspondences across sensory modalities or intermodal mappings that are stored in the brain and commingled with sensory-specific information such as touch, vision, and proprioception from the body. They are *analog representations*, body analogies if you will, for complex movement sequences. Third, these intermodal mappings are drawn from diverse sensory modalities as in something seen, heard or felt and fused with self-generated action because dance conveys aesthetic meaning through conceptual and sensory interplay.

There are six aesthetic mechanisms in movement: Imitation, style, visual resemblance, metaphor, metonymy, and statuses and roles. For example, a very young dancer might *imitate* an airplane or a bird in flight by flapping her arms. Likewise, when a dancer moves or behaves in a feline manner, she is demonstrating the capacity for *visual resemblance* or the outward form of movement. The illusion of overcoming gravity, however, through ballet pointe technique—as opposed to the use of heaviness of the lower body to convey pathos in modern dance—is an example of *style in dance* or the conventionalization of gestures and whole-body movements. The two tropes, metaphor and metonym, convey aesthetic meaning by *substituting one movement for another* as in placing the hands on each side of the head to express anxiety or using a part of a movement as a *metonym* to express a larger whole as in a war dance to convey the notion of impending battle. Lastly, *statuses* (a king) and *roles* (a romantic lover) are conveyed through, respectively, a courtly or coquettish dance. These aesthetic responses constitute embodied analogies to physical qualities, to the senses, and to the emotions. Rather than expressing actual sensory, physical or emotional experience, dance reveals aspects of reality, such as relationships among individuals, through embodied or gestic form.

Dance creates *gestic form* in two primary ways. Through *virtual self-expression* the dancer's concrete gestures capture the forms of feeling as in a hand outstretched to convey romantic desire. Through *logical expression* the dancer symbolizes concepts with the body as in the hands pressed together to one side of the head to suggest sleep. Both self-expression and logical expression are enacted through the interaction of bodily-gestural forms, the kinesthetic and projected space around the dancers' bodies, lighting, scenery, props, music, and audience reactions. For that reason, dance might be best understood as a field of virtual objects and forces including dancers, dance elements, space, time, and gravity. This is so because dance is governed by physical laws of motion. These include the physical laws of velocity, acceleration, position, and time as well as the relationship between these physical laws of motion and the forces that operate on things in motion. These physical laws create the illusion of virtual forces. For example, an *arabesque* in ballet where the dancer turns on one leg *on pointe* while the other is extended outward horizontally creates the illusion of the body suspended in space.

These *embodied analogies* occur both in the immediate present in a *body loop* in which occurrent physical, sensory, and emotional experiences assist physical movement and in an *as if loop* based on previous experiences stored in kinesthetic or long-term memory. A similar embodied process has been argued to underlie musical composition and improvisation. That is, the ability to integrate information across different sensory modalities enables individuals to perceive intensity, melody, spatial location, tempo, and rhythmic structure in music in an *amodal manner*, that is, to encounter it as one integrated experience.

Was dance the first art form originating in early *H. sapiens* as some scientists propose, or was there coeval emergence of dance, music, and visual art? While we won't be able to answer this question just yet, there is some interesting genetic data that picks out unique characteristics of professional dancers and dance ability. Genetic polymorphisms on genes SLC6A4 and AVPR1a are associated with aesthetic movement in professional dancers but not elite athletes, nonathletes or nondancers. Genetic polymorphisms are phenotypes (observable characteristics of someone such as their hair color) that reflect specific genetic causes from two or more alleles at the same gene locus on a chromosome. These differences in common genetic polymorphisms have been found to reflect variation in preferences, aptitude, and the need for expression through dance movement in individuals. The serotonin transporter gene (SLC6A4), for instance, not only regulates serotonin but also contributes to serotonin's effect in creating felt spiritual experience in susceptible individuals as suggested by dance's association with sacred rituals and ceremonies. The arginine vasopression receptor 1a gene (AVPR1a), on the other hand, appears to modulate affiliative behaviors (who you hang out with) as well as social communication (when, where, how, and who you communicate with). These results suggest that the association between aesthetic movement in professional dancers and genetic factors are modulated by aspects of personality including one's level of spirituality, capacity for empathy, need for social contact and interpersonal affiliation, as well as qualities of social communication.

Is There an Aesthetic *Sense* Underlying Music?

Like dance, discrimination and taste in music are, to a degree, faculties of the higher cognitive functions. One manner in which an aesthetic sense might function in music is through equivalencies to human movement. In his original contour theory, Peter Kivy, a philosopher of music, argued that the sonic shape of a musical piece was analogous to human emotional expression indicating a close association among music, the body, and human speech. This association is borne out by studies of contemporary English and French speakers and late nineteenth century and early twentieth century English and French composers. Measures of durational variability based on sampling of vowels and musical tones occur in the same general direction for both musical compositions and linguistic utterances taken from the same culture. Durational variability is the difference in duration between pairs of vowels or musical notes relative to the average length of the pair in a speech or musical sample. Vocal prosody, or intonation of a composer's native language, thus appears to influence the form or structure of a musical piece. This is so because composers most likely internalize rhythmic patterns of their native language very early in life and incorporate rhythms of folk and popular songs acquired in childhood in their mature compositions. In a similar manner, Stephen Davis, a music philosopher, contends that human emotions are presented directly in a musical work through analogy to human movement, behavior, physiognomy, the human voice, gait, and the like. According to Davis, music doesn't symbolize, describe or represent anything on its own, but like human movement, musical motion is inherent in the temporal dimension of music and turns on musical tension and relaxation. It is thus experienced as felt emotions in the listener.

Yet, aesthetic appreciation in music, as in dance, is not solely a mental activity but a form of cognition embodied. Empirical studies have demonstrated that when human infants are bounced on their parent's lap on either the second or third beat, they subsequently prefer the rhythmic pattern that matched their earlier experience. Thus, the experience of the infant's own body, but not the infant's self-initiated movement, is critical to the development of human musical understanding. This suggests that the effect is largely tied to the proprioceptive system rather than the vestibular system or vestibular-auditory interaction. Pedagogical practices in music including the Dalcroze, Kodaly, Orff, and Suzuki methods capitalize on the fact that basic elements of music such as rhythm and musical dynamics, intervallic relationships such as pitch and melody, and sonority can be most effectively taught through physical motion using such devices as rhythm, sight-singing or rhythmic solfege, and improvisation.

Is audition necessary for an aesthetic *sense* in music? Evelyn Glennie, a deaf percussionist, is able to play music through her ability to feel and adjust to the nuances of vibrations of musical instruments, the floorscape, and cognate materials. And there is evidence that the postlingually deaf, or individuals with acquired hearing loss, retain some musical imagery and appear to respond to musical sound. There are also anecdotal observations that composers who have lost some or all of their hearing are still able to compose, a few to astonishing heights, as in Ludwig von Beethoven in the last decade of his life. If not in music or dance, might there be an aesthetic *sense* in the visual arts?

Is There an Aesthetic *Sense* Underlying the Visual Arts?

The history and evolution of *image making* suggests that there is no aesthetic *sense* if one means by that a particular sensory modality necessary and sufficient for aesthetic enjoyment within a particular visual artistic genre. If the purpose of image making is to represent the sensible essence of objects as some artistic observers have asserted, then the form of the image is far removed from its actual appearance or any of its sensory properties or *qualia*. Rather, aesthetic appreciation and production emerge from the intellective imposition of meaning or semanticity on a set of marks or forms.

In human evolution, early hominins appear to have exploited three kinds of marks or forms: Color symbolism, decorative stylization, and direct marking. Meaning was initially impressed on shapes and forms preexisting in nature or from one's own body, although some have proposed that the origins of image-making may have been early hominins' own mental activity. These mental phenomena may have included *entopic phenomena* or sensations that seemingly arise from within the eyeball, such as afterimages and hallucinations, but actually emerge from physiological processes in the brain. In either case, Ernest Gombrich, the distinguished art historian, claims that the human mind originally imposed a few central narrative or classificatory schemes over the most recent ten millennia or so. For example, loss and renewal, and victory and defeat. These central schemes, he hypothesizes, arose from cultural developments in early human history rather than from any biological changes in the human brain. Such early image making may have served the storage and transfer of cultural knowledge, myth creation, social identity, food procurement, control of reproduction and fertility, as well as early forms of spirit worship. Indeed, of the 300 or so prehistoric caves that exist in Europe, anthropologists and other scientists have found evidence of many important aspects of modern visual art including perspective drawing, interlocking figures, surface preparation to control light and contrast, pointillism, false relief, the apparent capacity to switch easily from 2-dimensional to 3-dimensional imagery, and evidence of a social system of apprenticeship and training. Importantly, the most frequent subjects represented such as horses, bovines, and mammoths were dietarily the least important. Recent evidence, using uranium-thorium dating, indicates that an early hominin, *H. neandertalensis*, used perforated marine shells as ornaments and drew abstract images consisting of long lines, dot patterns, and the outline of the human hand, on the walls of caves dating back as much as 120,000 years ago long before modern humans, *H. sapiens* had settled in areas of eastern Spain where these artifacts were found. This represents the centrality of advanced cognitive capacities in very early humans in aesthetic production.

On the other hand, while infra-human primates (apes) may partake of visual and kinesthetic delight in applying paint to paper and create balanced and symmetrical designs, the marks appear to have no representational meaning. Yet, in young children, one can discern an incipient aesthetics in their ability to comprehend, for instance, simple stylistic differences in groups of paintings from different historical eras, a noteworthy cognitive ability given little or no prior experience.

The empirical evidence indicates that aesthetic perception in the visual arts is modular with perceptual systems in the brain functionally specialized for the recognition of color, form, motion, direction and orientation, faces and facial expression, and body language. Color vision itself, however, is far removed from actual appearance in the real world, because for one, the human visual system only perceives light within a very narrow range of the electromagnetic spectrum (400–780 nm). Color consistency and color unity across an object's surface—color and form are inextricably linked because object borders are instrumental in the creation of shape and form—are computed by visual areas of the brain based on a simultaneous comparison of a color patch on the surface of an object and its immediate visual surround. Both are significantly influenced by judgment, learning, and memory since there are both natural and unnatural colors. Individuals missing the genes for red, green or blue color perception, are either red-green or blue-yellow colorblind or both and the red and green color genes are sex-linked and more common in males. There is some evidence that certain females may compute color with an additional *cone*—cells on the back of the retina that sense color—in the red range of the electromagnetic spectrum potentially heightening their aesthetic appreciation. Conversely, individuals who have lost visual sensation rely on other sensory systems as the following excerpts from a blind Professor of Religious Studies attest.

> A blind person is simply someone in whom the specialist function of sight is now devolved upon the whole body, and no longer specialized in a particular organ. For instance, rain has a way of bringing out the contours of everything; it throws a colored blanket over previously invisible things; instead of an intermittent and thus fragmented world, the steadily falling rain creates continuity of acoustic experience. **Facial vision** [my emphasis] is the sense of pressure on the skin of the face rather than or upon the ears.[5]

Certainly, the sensory impressions of temperature and pressure on the face and bodily surfaces and sound pressure upon the tympanic membranes of the inner ear as well as the analysis and discrimination of sound have always assisted the sensorium of the blind. Their investigation, in fact, originally commenced with the French philosopher, Denis Diderot's early studies and prognostications on the subject in the 1700s as I mentioned earlier.

Besides color, luminosity or perceived brightness is an important aesthetic value because luminance of color differs depending on ambient light and influences the perception and shape of objects. Rembrandt exploited this aesthetic feature extensively in his artistic work by manipulating the luminance contrast between the background and subject of his paintings, a technique called *countershading*. He used it to great effect in his 1632 work, "Meditating Philosopher," predominantly through the use of chiaroscuro or tonal contrast between light and dark that creates three-dimensional form, stereopsis, and abets color mixing. The striking result that Jean-Auguste Dominique Ingres achieved two centuries later in his portrait of "Princess Alabert de Broglie" foreshadowed the realism of early photography.

Familiarity, context, and typicality are also important aesthetic values and along with color and luminosity play a significant role in aesthetic visual perception even though immediate perception itself (color, luminosity) is informationally encapsulated and less porous to conscious knowledge and past experience, of which these three incorporate. *Familiarity*, of course, refers to acquaintance with an aesthetic object or event in a painting or print such as an apple or pear or an event like war or the gathering of citizens in a town square. Related to familiarity is *typicality* or whether these objects or events are conventional in pictorial depiction or not. *Context* alludes to the relationship of the objects and events to each other and the overall scene denoted in the depiction. Nonetheless, discrepancies among vision, touch, and proprioception—what's been called the *unity of the senses*—are re-calibrated by central cognitive systems when there is conflicting information among these various sensory modalities. But conflicts between visual memories and conscious, objective knowledge are typically resolved in favor of the former even though this stored knowledge is continuously compared to ongoing perception with all of its inherent conflicts.

The same is true of visual ambiguity. If the visual stimulus remains constant but the percept changes—as in ambiguous figures such as a line drawing switching back and forth between a young girl and old women—higher brain centers, the fronto-parietal cortices, kick in as we become aware of the discrepancy and voluntarily switch visually between alternative visual percepts. It is this ability to experience multiple perceptual viewpoints from a work of art, not just in patently ambiguous figures, but in all visual art that renders aesthetic pleasure according to one of its most prominent proponents, the neuroscientist, Semir Zeki.

As it turns out, the experience of *beauty* in art has a significant neuropsychological component. A network of brain regions including the medial orbito-frontal cortex, anterior cingulate, left parietal cortex, and motor cortex appear to underlie the perception of beauty. The medial orbito-frontal cortex along with important linkages to the corpus striatum, a subcortical structure, is involved in the perception of rewarding stimuli and is activated by novelty and salience, and thus appears vivid to the senses. This cortical-subcortical network contains neurons that respond to food and drink, sexual arousal, psychoactive drugs such as nicotine, and monetary reward. The anterior cingulate cortex is a deep cortical structure central to attention as well as emotional arousal and perception. The ventral part of the anterior cingulate cortex is activated by the emotions of anxiety and sadness and appears to be involved with empathic awareness and the appreciation of music. The dorsal part is linked to social communication areas in the brain such as theory of mind regions and is activated by, among other things, romantic love. The left parietal cortex undergirds spatial awareness, undoubtedly a prominent feature of visual aesthetics. The motor cortex is involved in approach and withdrawal from aversive and rewarding stimuli along with areas in the right hemisphere and the amygdala that play a role in the perception of negative emotions. It appears, then, that there is not a separate area in the brain for the perception of *beauty and its opposite* as both appear to be mediated by the same underlying neural network.

This brings us to the issue of taste and discrimination. David Hume believed that taste was a human propensity that cut across various forms of aesthetic pleasure whereas Immanuel Kant elevated aesthetic enjoyment to a separate realm supervenient upon bodily ones. Others have proposed that aesthetic appreciation is a universal human property as supported by the aforementioned neurological evidence. Indeed, Kant acknowledged that aesthetic experience and the experience of beauty were universal because they were undergirded by mental processes that were themselves ubiquitous and present in all modern humans.

The Aesthetic *Sense* in Other Sensory Modalities

When discussing the aesthetic enjoyment of wine, for instance, master sommeliers and viticulturists emphasize the acquisition of the ability to discriminate such aesthetic elements as the body of the wine (light, medium or full body); dryness or the lack of sweetness in a wine; acidity or crispness; as well as oakiness or the addition of certain flavors within the barrel in which the wine is aged. Additionally, there are such aesthetic components as the wine's structure or the combination of body and tannin in a wine, and the floral or fruit quality of a wine such as its butteriness (due to a trace component called diacetyl), herbaceousness or the prominence of a grassy taste, spiciness (*sweet* as in cinnamon, ginger, cloves, anise, licorice, nutmeg, and allspice or *savory* as in black pepper, white pepper, cumin, coriander, and mustard powder), as well as floral scents such as orange blossom, honeysuckle, lilac, lavender, lily, hyacinth, jasmine, and hibiscus, to name a few. These abilities, moreover, do not just rely on taste receptors on the surface of the tongue or *glossa*, but may be superceded by olfactory receptors in the nasal cavities (the wine's nose), visual appearance (the wine's color and clarity) as well as the haptic sensation of the wine on the interior surfaces of the mouth as in assessing the wine's concentration of tannins. Aesthetic enjoyment is most likely further compounded by vestibular and proprioceptive feedback in the holding as well as their initial design of the various wine glasses and decanters that shape and modify the aesthetic pleasure conveyed by the external sensory apparatus. The olfactory system, moreover, is importantly connected to multimodal areas in the orbitofrontal cortex that integrate information from the gustatory, olfactory, affective, and intellective areas of the brain, thus providing a holistic experience of food and drink.

The cognitive complexity of acquiring such an aesthetic ability is compounded by learning which foods and which wines go best together. Classic pairings—known to whet the appetites of gastronomes, at least in Western cultures—include such delicacies as steak and full-bodied reds (Cabernet, Merlot), champagne and caviar, goat cheese and Sancerre (a dry, herbaceous white wine), chips and salsa combined with a sweet Riesling, and so on. The justification for such pairings is that they are acquired higher-order cognitive abilities. Such abilities, of course, categorize and classify things in the world based on the sensory attributes of objects and such aesthetic pleasures are probably universal across human cultures although obviously not specific cultural objects nor the pairings themselves.

Is there an aesthetic *sense*? Not if what is meant by an aesthetic *sense* is some property of sensation that situates aesthetic appreciation and understanding in a particular form of sensory perception or some truncated collection of sensory modalities. Rather, aesthetic appreciation and understanding are the operation of the higher cognitive functions in close communion with the body. These higher cognitive functions are reliant on the senses as well as the body and importantly shaped by culture, experience, and informal and formal education. From the heights of neuroscience, beauty is the brain's response to rewarding stimuli that activate areas of the brain that respond to novelty and salience. These rewarding stimuli cut across classes of social and nonsocial objects (people and things) and include, not just the perception of persons and bodies, but food and drink, music, the visual arts, aesthetic movement, and related areas. As a consequence, what we know about the properties of brains and the properties of bodies suggests that an *aesthetic sense* is universal in human experience. That is, all individuals in all cultures can potentially appreciate the beauty of a sunrise or sunset or the beauty of music or dance or visual art, because they are undergirded by minds, brains, and bodies that are ubiquitous and universal features of human understanding.

With that in mind, we look next at the embodied nature of personality.

Notes

1. The quote, "we have discovered nothing new ..." is from R. White, *Prehistoric Art: The Symbolic Journey of Humankind*, 2003.
2. The quote, "stable cultural realm that preserves ..." is from M. Csikszentmihalyi, *Society, Culture, and Person: A Systems View of Creativity*, 1988.
3. The quote, "select from variations produced by individuals those worth preserving" is from M. Csikszentmihalyi, *Society, Culture, and Person: A Systems View of Creativity*, 1988.
4. The quote, "Ian's enjoyment of any kinetic melody ..." is from Jonathan Cole & Ian Waterman, *Pride and a Daily Marathon*, 1995.
5. The quote, "A blind person is simply someone ..." is from J. M. Hull, *Touching the Rock: An Experience of Blindness*, 1990.

Bibliography

Albers, J. (1963). *Interaction of color*. New Haven: Yale University Press.

Bachner-Melman, R., Dina, C., Zohar, A. H., Constantini, N., Lerer, E., Hoch, S., Ebstein, R. P. (2005, September). AVPR1a and SLC6A4 gene polymorphisms are associated with creative dance performance. *PLoS Genetics*, *1*(3), 0394–0403.

Bartels, A., & Zeki, S. (2004). The neural correlates of maternal and romantic love. *NeuroImage*, 21, 1155–1166.

Bastianich, J., & Lynch, D. (2005). *Vino Italiano: The regional wines of Italy* (Rev. ed.). New York, NY: Clarkson Potter.

Chekhov, M. (1991). *On the technique of acting*. New York, NY: HarperCollins.

Clynes, M. (1989). *Sentics: The touch of emotions* (Rev. ed.). Dorset, England: Prism Press.

Cole, J., & Waterman, I. (1995). *Pride and a daily marathon*. Cambridge, MA: MIT Press.

Cuthill, I. C., Allen, W. L., Arbuckle, K., Caspers, B., Chaplin, G., Hauber M. E. ... Caro, T. (2017). The biology of color. *Science*, *357*(6350), 470–477.

Csikszentmihalyi, M. (1988). Society, culture, and person: A systems view of creativity. In R. J. Sternberg (Ed.), *The nature of creativity: Contemporary psychological perspectives* (pp. 362–385). New York: Cambridge University Press.

Davies, S. (1994). *Musical meaning and expression*. Ithaca: Cornell University Press.

Davis, W. (1986). The origins of image making. *Current Anthropology*, *27*(3), 193–215.

de Waal, F. B. M. (1999, November 19). Apes with an oeuvre. *The Chronicle of Higher Education*, pp. B1–B6.

Donald, M. (1991). *Origins of the modern mind: Three stages in the evolution of culture and cognition*. Cambridge, MA: Harvard University Press.

Dutton, D. (2005). Aesthetic universals. In B. Gaut & D. M. Lopes (Eds.), *The Routledge companion to aesthetics* (pp. 279–291). (2nd ed.). London: Routledge.

Faas, E. (2002). *The genealogy of aesthetics*. Cambridge: Cambridge University Press.

Gardner, H. (1990). *Art education and human development (Occasional Paper Series, Vol. 3)*. Los Angeles, CA: Getty Publications.

Gaut, B., & Lopes, D. M. (2005). (Eds.). *The Routledge companion to aesthetics* (2nd ed.). London: Routledge.

Glennie, E. (2018). *Evelyn Glennie home page* accessed 17 March 2018. Available: https://www.evelyn.co.uk.

Gombrich, E. H. (1963). *Meditations on a hobby horse and other essays on the theory of art.* London: Phaidon Press.

Gombrich, E. H. (1969). *Art and illusion: A study in the psychology of pictorial perception* (3rd ed.). Princeton: Princeton University Press.

Goodman, N. (1976). *Languages of art: An approach to a theory of symbols* (2nd ed.). Indianapolis: Hackett.

Graziano, A. B., Peterson, M., & Shaw, G. L. (1999). Enhanced learning of proportional math through music training and spatial-temporal training. *Neurological Research, 21*(2), 139–152.

Habibi, A. & Damasio, A. (2014). Music, feelings, and the human brain. *Psychomusicology: Music, Mind, and Brain, 24* (1), 92–102.

Hull, J. M. (1990). *Touching the rock: An experience of blindness.* NY: Random House.

Hume, D. (1998). Of the standard of taste. In S. Copley & A. Edgar (Eds.), *Selected essays* (pp. 133–153). Oxford: Oxford University Press (Originally published 1757).

Hurley, S., & Chater, N. (2005). *Perspective on imitation: From neuroscience to social science.* Cambridge, MA: MIT Press.

Iacoboni, M. (2005). Neural mechanisms of imitation. *Current Opinion in Neurobiology*, 15(6), 632–637.

Iacoboni, M., Woods, R. P., Brass, M., Bekkering, H., Mazziotta, J. C., & Rizzolatti, G. (1999). Cortical mechanisms of human imitation. *Science*, 286(5449), 2526–2528.

Jaques-Dalcroze, E. (1967). *Rhythm, music, and education* (Rev. ed., trans., H. F. Rubenstein). London: The Dalcroze Society (Original work published 1921).

Jaques-Dalcroze, E. (1976). *Eurhythmics, art, and education* (C. Cox, Ed. and F. Rothwell, Trans.). NY: Arno Press (Original work published 1930).

Jones, S. (2005). The role of mirror neurons in imitation. In S. Hurley & N. Chater (Eds.), *Perspectives on imitation: From neuroscience to social science*. Vol. 1. Mechanisms of imitation and imitation in animals. Cambridge, MA: MIT Press.

Kant, I. (1764). *Observations on the feeling of the beautiful and the sublime* (J. T. Goldthwait, Trans.). Berkeley: University of California Press.

Kant, I. (1790). *The critique of judgment* (J. C. Meredith, Trans.). Oxford: Clarendon Press.

Kawabata, H., & Zeki, S. (2004). Neural correlates of beauty. *Journal of Neurophysiology*, *91*(4), 1699–1705.

Kivy, P. (2002). *Introduction to a philosophy of music.* Oxford: Clarendon Press.

Korsmeyer, C. (2005). Taste. In B. Gaut & D. M. Lopes (Eds.), *The Routledge companion to aesthetics* (pp. 267–277). (2nd ed.). London: Routledge.

Kosslyn, S. M. (1980). *Image and mind.* Cambridge, MA: Harvard University Press.

Langer, S. K. (1953). *Feeling and form: A theory of art.* New York, NY: Scribners.

Larson, G. O. (1997). *American canvas: An arts legacy for our communities.* Washington, D.C.: National Endowment for the Arts.

Laban, R. (1966). *Choreutics* (L. Ullmann, Ed.). London: MacDonald & Evans.
Laws, K. (2005). *Physics and the art of dance: Understanding movement*. Oxford: Oxford University Press (Originally published in 1984).
Lewis-Williams, D. (2002). *The mind in the cave: Consciousness and the origins of art*. London: Thames & Hudson.
Livingstone, M. (2002). *Vision and art: The biology of seeing*. New York, NY: Harry N. Abrams.
McMahon, J. A. (2005). Beauty. In B. Gaut & D. M. Lopes (Eds.), *The Routledge companion to aesthetics* (2nd ed.). (pp. 307–319). London: Routledge.
McNeill, W. H. (1995). *Keeping together in time: Dance and drill in human history*. Cambridge, MA: Harvard University Press.
Melser, D. (2004). *The act of thinking*. Cambridge, MA: MIT Press.
Montero, B. (2006). Proprioception as an aesthetic sense. *The Journal of Aesthetics and Art Criticism*, *64*(2), 231–242.
Noe, A. (2004). *Action in perception*. Cambridge, MA: MIT Press.
Patel, A. D., & Daniele, J. R. (2003). An empirical comparison of rhythm in language and music. *Cognition*, *87*(1), B35–B45.
Phillips-Silver, J., & Trainor, L. J. (2005). Feeling the beat: Movement influences infant rhythm perception, *Science*, *308*(5727), 1430.
Rauscher, F. H., Shaw, G. L., Levine, L. J., Wright, E. L., Dennis, W. R., & Newcomb, R. L. (1997). Music training causes long-term enhancement of preschool children's spatial-temporal reasoning. *Neurological Research*, *19*(1), 2–8.
Riviere, J. (2002). Present tendencies in painting. In C. Harrison & P. Wood (Eds.), *Art in theory, 1900–2000: An anthology of changing ideas* (pp. 183–187). Oxford, UK: Blackwell (Originally published 1912).
Robinson, A. I. (2005). *Great wine made simple: Straight talk from a master sommelier*. New York, NY: Broadway Books.
Seitz, J. A. (1987, February). Unpublished interview with David Wolfe, choreographer, Spoke the Hub Dancing, Brooklyn.
Seitz, J. A. (1993, March/April). I move … therefore I am. *Psychology Today*, *26*(2), 50–55.
Seitz, J. A. (1994, August). *The development of aesthetic movement skills*. Paper presented at the meeting of the International Association for Empirical Aesthetics, Montreal, Canada.
Seitz, J. A. (1996). The development of aesthetic movement: Linkages to preschool education. *Journal of Early Education and Family Review*, 5(3), 7–9. (Excerpted on the ERIC database at http://ericae.net/ericdb/EJ531338.htm).
Seitz, J. A. (1999). Political science and creativity. In M. A. Runco & S. Pritzker (Eds.), *The Encyclopedia of Creativity* (Vol. 2; pp. 417–421). San Diego, CA: Academic Press.
Seitz, J. A. (2001a). A cognitive-perceptual analysis of projective tests. *Perceptual and Motor Skills*, 93(2), 505–522.
Seitz, J. A. (2001b). Mind, musical improvisation, and the body: The microgenesis of jazz improvisation. In J. A. Seitz (Chair), *Adaptive and maladaptive forms of embodied cognition*. Paper presented at the 31st Annual Symposium of the Jean Piaget Society, Berkeley, CA.
Seitz, J. A. (2002). Mind, dance, and pedagogy. *Journal of Aesthetic Education*, 36(4), 37–42.

Seitz, J. A. (2003a). The political economy of creativity. *Creativity Research Journal, 15*(4), 385–392.

Seitz, J. A. (2003b). A communitarian approach to creativity. *Mind, Culture, and Activity: An International Journal,* 10(3), 243–247.

Seitz, J. A. (2005). Dalcroze, the body, movement, and musicality. *Psychology of Music, 33*(4), 419–435.

Shepard, R. N. (1978). Externalization of mental images and the act of creation. In B. S. Randhawa & W. E. Coffman (Eds.), *Visual learning, thinking, and communication.* New York, NY: Academic Press.

Smetacek, V., & Mechsner, F. (2004). Making sense. *Nature, 432*(7013), 21.

Smith, E. E., & Jonides, J. (1999). Storage and executive processes in the frontal lobes. *Science, 283*(5408), 1657–1661.

Supa, M., Cotzin, M., & Dallenbach, K. M. (1944). "Facial vision": The perception of obstacles by the blind. *The American Journal of Psychology, 57,* 133–183.

Sussman, A., & Hollander, J. B. (2015). *Cognitive architecture: Designing for how we respond to the built environment.* New York, NY: Routledge.

Vialou, D. (1996). *Prehistoric art and civilization.* New York, NY: Henry N. Abrams.

White, R. (1994). Les archives Paléolithiques. *La Recherche, 25,* 746–750.

White, R. (2003). *Prehistoric art: The symbolic journey of humankind.* New York, NY: Harry N. Abrams.

Zeki, S. (1999). *Inner vision: An exploration of art and the brain.* Oxford: Oxford University Press.

Zeki, S. (2004). The neurology of ambiguity. *Consciousness and Cognition, 13,* 173–196.

· 8 ·

PERSONALITY

What Is Personality?

> The dynamic organization within the individual of those psychophysical systems that determine his characteristic behavior and thought.[1]

Personality has been conceptualized through a number of theoretical lenses including enduring dispositions or traits that direct behavior; motivations, defenses, and conflicts that influence personality formation; the subjective experience of oneself; the role of behaviors and emotions in constructing personality; the formation of personality through relations with others; and the shaping of dispositions by way of the genes, the brain, and the body.

Dispositional View

The received view in traditional theories of selfhood and personality assumes that self and personal identity are thoroughly mind-based. Take, for instance, the dispositional view of human personality. One major view of personality informed by the dispositional view is the *Big Five factor model.* It consists of the following five primary personality traits and their opposites that are believed to encompass the majority of human personalities: Neuroticism, extraversion, openness to experience, agreeableness, and conscientiousness. *Neuroticism* is a much-maligned term, but it simply means a tendency to foreground negative emotions in one's overt behavior with others such as projecting anxiety or insecurity. *Extroversion* refers to being outgoing and hence friendly and spontaneous. Independent and artistic might be dispositional terms associated with *openness to experience*. An *agreeable person* would be someone who evinces good-naturedness and a helpful disposition. A goal-directed and careful person would be someone who possesses the trait of *conscientiousness*.

The problem with the dispositional view is that people act differently with sundry types of people as well as in diverse situations and contexts. An extroverted person might be quiet and unassuming on a blind date whereas an introverted person might be quite outgoing. Similarly, their personality dispositions may change with *situation* such as a business meeting or family gathering or with the *context* such as a romantic proposal or passing of a close friend. This particular view of human personality has a long history going back to the Greeks. They believed in four fundamental temperaments: *Choleric* or irritable and strident, *melancholic* or sad and depressed, *sanguine* or cheery and optimistic, and *phlegmatic* or calm and restrained. In the 1950s, two cardiologists suggested, after careful study of men in their 30s, 40s, and 60s, that a strong drive to succeed and manage their time rigorously could lead to a *Type A personality* and increase the chances of cardiovascular disease. The *Type B personality*, however, was less prone to stress, and less career and success driven, and thus less likely to develop disease. Some observers of human personality have posited that there are really two major kinds of traits: *Surface* and *source* traits. The former comprises individuals possessing such surface characteristics as thriftiness, foolishness or curiosity while the latter includes dispositional dimensions such as being dominant or submissive in interactions with others.

In first decade of the 21st century, several researchers advocated for including the *honesty-humility dimension* in the Big Five personality dimensions. This trait is concerned with an individual's sincerity, sense of fairness, and lack of a propensity to be greedy. It is also negatively correlated with the *dark triad* of personality traits including Machiavellianism, psychopathy, and narcissism. Possessing a Machiavellian trait indicates someone who is manipulative and deceitful. A psychopathic trait bespeaks an individual who lacks empathy for others and is antisocial. Narcissism adds a dimension of strong self-centeredness. Interestingly, these traits are devoid of situation or context nor do they change when interacting with different kinds of individuals. They are posited to be universal and stable across individual interactions and sociocultural contexts. What is missing out of these accounts is the role that the *mind-brain-body* (MBB) plays in the construction of our sense of self and personality.

The Role of the Mind-Brain-Body (MBB)

As I discussed previously, there is evidence for a distributed neural network including the somatosensory cortex, limbic system, and cortical regions central to object- and self-recognition, that forms the neural basis for corporeal awareness, that is, one's representation of one's body or body schema. This body schema changes over time and, as a result, one's personal identity is modulated by body-centered aspects of the mind-brain-body (MBB). That is to say, one's personal identity and sense of self are constructed from body-centered aspects of the MBB and are modulated over time by changes to one's body schema.

Access to one's emotions, the ability to symbolically elaborate on one's feelings, the skill in using cognitive-affective states to guide one's behavior, and the capacity to read intentions and influence others constitutes a working definition of personal identity and one's sense of self. However, personal identity is not solely mind-based. As I discussed earlier, the body influences personality by tagging affective states with somatic markers. A somatic marker is a visceral or nonvisceral sensation that marks a cognitive-affective state. Thus, basic bodily regulatory systems influence conscious, cognitive processes. Subjectivity arises from changes that take place in bodily states. That is, subjective awareness of self may arise by way of the body's awareness of itself through interoceptive, exteroceptive, and proprioceptive senses as discussed in Chapter 1.

From the study of the brain's prefrontal cortices, three sets of neural structures have been identified and are hypothesized to underlie this sense of self, discussed in Chapter 5: Sensory cortices (object representation), sensorimotor cortices (self-representation), and subcortical structures such as the cerebellum, basal ganglia, and thalamus that create representations of the self, separate from the self that is responding to the environment. Since these linked structures are nonverbal and constitute bodily-based aspects of the MBB they might also occur in infrahuman primates and other species.

Reciprocal and parallel neural pathways between the cerebellum, traditionally viewed as controlling gross and fine motor functions but hypothesized to play a much larger role in thought itself, and the frontal cortex where working memory, executive functions, and aspects of one's sense of self are housed, suggest that the MBB is intimately involved in aspects of personality. For instance, as discussed earlier, there is evidence that inadequate network connections among the dorsolateral prefrontal cortex, thalamic pathways, and cerebellar sites leads to *cognitive dysmetria* or problems in the processing, coordination, and prioritizing of information that may play a role in the genesis of schizophrenia and other disorders of personal identity. Similarly, the *cerebellar cognitive affective syndrome* is linked to disturbances in language and personality because of disruption of these network connections.

While traditional theories of selfhood postulate that personality is wholly mind-based, new views of the brain and the role of sensorimotor capacities in human cognition, suggest that one's sense of self and personal identity are constructed from body-centered aspects of the MBB that are modulated by changes to one's body schema over time. Indeed, case studies of the breakdown of personal identity as a result of mental illness and after brain damage appear to support this view.

Borderline Personality Disorder (BPD)

Borderline personality disorder (BPD) is a disorder of interpersonal relationships that commences in early adulthood and is frequently found in females. It is marked by instability in relations with others that affects one's emotional self, one's self-image, and results in frequent and sustained impulsive behaviors. The emotional system in BPD appears overly labile and reactive and individuals may engage in self-injurious and parasuicidal behavior and, as a result, make frequent attempts to harm themselves. There is impoverished understanding of others and often a sense of depersonalization from the self and the body as well as a dissociation from reality itself. This is because of a weak core self and experience that accompanies it that often consists of shameful feelings. Bipolar disorders (cycles of mania and depression) are common along with unipolar disorders such as depression or its lesser form, dysthymia.

In young adults with borderline personality disorder, there is typically pervasive and persistent stress from childhood that may include a history of sexual abuse, temper tantrums, and poor sleep hygiene including frequent rocking in bed. These life experiences negatively affect the limbic system and transform it in detrimental ways including maladaptive mechanisms of emotional sensitivity and arousal.

Case Study #2

D. H. is a 50-year-old adult female with mild mental retardation and a history of cerebral stroke and bipolar disorder. Psychometric intelligence—what standardized intelligence tests measure—is 62 and more than two standard deviations below the mean of 100 (the arithmetic average), which partly accounts for her mental challenges. Linguistic and visuospatial abilities are weak. She is currently taking antipsychotic and anticonvulsant medication typically prescribed in individuals with mood disorders and an anticholinergic drug to offset the effects of the antipsychotic medication.

There is ample eye contact, but she is lethargic, behaviorally inhibited, and fearful although she smiles on occasion. Affective range and intensity—the range of emotions and the intensity of their expression—is limited. There is some evidence of computational competencies as she can count, add, and subtract. There is evidence of self-awareness as well as the capacity for joint attention or the ability to share in a common interest or activity. She speaks in multiword sentences with some syntactic, pragmatic, and semantic competence. That is, she demonstrates some capacity to arrange words in their correct grammatical order, understand another's intended meaning, as well as speak in meaningful sentences. On a measure of lexical retrieval (retrieving words from memory) she easily names pictured objects. She is adept at memorizing and repeating back sentences on a task of working memory. Working memory is a measure of one's ability to temporarily hold information in memory and process and think about it. She is unable, however, to remember more than a 3-digit string of numbers (8–7–9). Memory for previously learned verbal lists (robin, carrot, piano, and green) is well below the average range of normal variation for her chronological age as she needs extensive prompting on 75% of the initial items or she forgets the others. She has poor recall across three consecutive trials for a 12-word list drawn from the semantic categories of occupations, sports, and vegetables. This is known as categorical recall. Overall, working memory is below the average range of normal variation for both verbal and numerical items. Nonetheless, on a measure of verbal analogical reasoning she was able to place simple verbal concepts ("corkscrew," "hammer") within their correct superordinate categories ("tools") but not for more complex verbal relations.

She describes the activities of various human figures in a line drawing but with limited insight. She displays poor judgment and reasoning to common everyday ethical dilemmas. She is able to write her full name and read the headlines of a newspaper indicating that she is functionally literate, that is, possesses the ability to read on a fourth-grade reading level or above. She possessed adequate fine, but not gross motor abilities. That is, she is competent with her hands and expresses adequate fine-motor skills in writing tasks, but she is not fluid in whole-body movements such as jumping, skipping or running.

The entire mind-brain-body axis of D. H. is affected by this multifaceted case of bipolar disorder and poor intellectual and interpersonal functioning. Motor functioning, cognition, language, memory, interpersonal interactions, nonverbal behavior, and social interaction are all equally affected in some manner.

Body Image and Body Schema

Shaun Gallagher, a philosopher, we discussed earlier alleges that body image and body schema are actually two different notions. Whereas a body image makes use of visual imagery in consciousness and is distinct from other objects in the environment, a body schema is indistinct from the environment in which it inhabits and exists non-consciously in our mind-brain-body.

The term, "body schema," originated with the British neurologist, Henry Head, who first used this designation in 1911. It has been subsequently defined and re-defined as an assemblage of processes of the MBB that continually register the various parts of the body in space and spatially organize its action. It is informed by sensorimotor and proprioceptive information from our body as well as diverse regions of the brain. It can be affected by disease states such as deafferentation that we saw in the case of Ian Waterman where there is selective loss of proprioception as well as in the phenomenon of phantom limbs. In both cases, our sense of our body schema is radically changed.

On the other hand, *body image* is heavily influenced by our mood states as depressed individuals characteristically negatively distort their body images and see themselves as less attractive than non-depressed individuals. This is so even when independent raters judge them as equally attractive. On the other hand, non-depressed individuals positively distort their body images and see themselves as more attractive than they really are, a form of self-deception used to enhance emotional well-being. We see the same bias between the genders. Women tend to distort their perception of their bodies in a negative direction whereas men are disposed to misrepresent their perceptions in a positive direction and in an often, self-aggrandizing manner. To put it another way, men tend to be satisfied with their body image and believe that women feel the same. Women tend to be less satisfied with their body image and believe men want them to be often thinner than they actually are.

Mentalizing Network

Let's look at the *mentalizing system or network* for a moment, that is, the ability to understand one's own and other's behavior. The mentalizing system has a number of interesting properties. Like my early research on physiognomic metaphor I discussed in Chapter 3, the ability to discriminate between inanimate and animate objects is vastly important or otherwise you might end up attributing mental states to your pet rock. Second, to follow the gaze of another person is consequential in recognizing the attention of another and any action they may take in the direction of that gaze such as acknowledging a mutual friend or an appetizing food source. Despite all of this, the end point of the mentalizing system is the ability to separate out the actions of others from your own. Like many other abilities in human and primate evolution, these abilities did not arise *de novo* but were *exapted* from preexisting ones. The mentalizing system appears to have evolved from the *dorsal action* or *where system* rather than from the object identification system or the *ventral what system*. Prior to the emergence of Great Apes or hominids (chimps, bonobos, humans, gorillas, and orangutans), the ability to discern small differences in facial expression, tell individuals apart, and understand social relations and dominance hierarchies was tantamount to survival in any highly social species. It relied on the *ventral what system* to identify crucial differences in object qualities. But with the emergence of the mentalizing system in some higher primates and early hominins as well as probably other species, the capacity for comprehending actions and their goals in oneself and others, as well as the intentions behind them, became important in ever-expanding cultural groupings and the close quarters they tended to inhabit. So, the brain evolved to encompass areas of the posterior superior temporal sulcus (STS) for noting people, actions, and their outcomes as this region is activated by *biological motion*; inferior parts of the frontal cortex for comprehending their *actions and goals*; and the anterior cingulate and medial prefrontal cortices for *understanding the mental states of oneself and others*. It appears, then, that the *mentalizing system* evolved from an earlier system for *representing actions*.

Autism, of course, is a neurodevelopmental impairment of the capacity for understanding others. *Alexithymia*, however, is often defined as an inherent inability to decode feelings in oneself. We looked earlier at the causes and consequences of autism in the cerebellum and lower brain regions. Alexithymia is thought to arise, however, from genetic and neural causes originating in the various higher cortices. It may be due to dysfunction of the anterior cingulate cortex and thus interfere with accessing emotional states in oneself. Both autism and alexithymia profoundly affect the normal operation of this mentalizing system.

On the other hand, *neurodegenerative disorders* in mid- and late-age adults such as frontotemporal dementia (FTD), result in impairments in impulse control and social awareness due to degeneration of frontal and temporal regions of the brain, which are intrinsic and instrumental to the operation of the mentalizing system. Typically, in FTD there is lack of empathic feeling, heightened perseveration such as repeatedly saying or doing the same thing, and a lack of restraint in following everyday social conventions. There are problems in planning and reasoning as well as understanding another's point of view that is often referred to as lack of executive control and thus comprising a *dysexecutive syndrome*. Early primitive reflexes manifested in development during infancy are often expressed in advanced FTD due to the lack of inhibition by the frontal lobes. These rudimentary reflexes include, among other ones, the palmar grasp (grabbing at objects) and the glabellar reflex (blinking in response to being tapped on the forehead).

One central aspect of personality and dispositions is, of course, the underlying emotions that support people's selves and their expression. Let's look at the importance of the emotions in our everyday lives by way of another thought experiment.

A Thought Experiment: A Creature Without Emotions 2.0

Imagine a creature or some kind of unusual animal without any emotional life. Could such a being survive in the everyday world?

The Harvard philosopher, Nelson Goodman, once remarked that we probably rely more on feelings in our daily lives than arid and cold cognition. Indeed, much of everyday thought is bound up with our fluctuating emotional lives. For instance, as some clinicians are fond to point out, strong emotional arousal has a tendency to significantly narrow our range of attention. Attention itself is a conscious process in which we selectively focus our mind-brain-body (MBB) on one thing while we ignore other things, divide our attention among several things, or sustain our attention on a single thing over an extended duration of time. What we focus our minds and bodies on and what we think about is highly determined by our level of emotional arousal as well as the range and kinds of emotional states we have experienced in the past.

Antonio Damasio, a neurologist, suggests that bodily signals—visceral sensation emanating from our abdominal area as well as bodily sensations arising from other parts of the body—modify the way the brain handles information, and this is accomplished primarily through use of our prefrontal cortices in the thinking part of our brains. Because they lack these bodily signals or somatic markers patients with damage to the frontal lobes are unable to even anticipate the future. That is, they are unable to predict future events, such as make simple, ordinary choices in everyday life, because they possess no biasing somatic state to rely on in decision-making.

Imagine yourself standing in a buffet line and having to decide between a piece of chocolate cake and a piece of apple pie. How does anyone typically make this type of decision? Certainly, there is no logical choice between the two if they are essentially the same price or calories. That is, there is no logical calculus that one can apply to arrive at deciding whether one option or another is preferable. That's because we typically make choices based on our preferences and our preferences are closely aligned with our pleasurable or non-pleasurable experiences in past encounters. We choose the chocolate cake because of the extraordinary memories we experienced of its exuberant sweetness, notes of cheery and almond, and the dark, rich flavors underneath—or something similar. The sensual joy of a gourmand is what prepared us for this experience not a series of logical steps in deductive thought.

It would appear, then, that a creature without emotions would thus be helpless in the most mundane of everyday tasks. Captain Spock of Star Trek fame is a convenient dramatic fiction not a character that has any rational basis in science or real life. Just like children that fail to survive because of a lack of ability to perceive and respond to pain (commonly referred to as *congenital analgesia* denoting a class of sensory neuropathies), our fictional Spock, bereft of the ability to perceive and respond to emotions in himself and others, would be unable to make even the simplest of everyday decisions—often to avoid harm—or form enduring relationships with caregivers and others that are essential for survival.

Q.E.D.: "Spock" then is just a dramatic fiction.

Without an emotional life, then, dispositions and traits would be devoid of meaning. For one thing, emotions are gateways to understanding the motivations behind actions. Emotion and motivation are closely linked, and they are expressed through movement, action, and activity. You fathom the disposition of someone who is introverted by not just what they say, but how they say it and by their actions and activity through facial expressions and vocal prosody, whole-body movements, and even the proxemics of how they align their bodies with someone when interacting and how close they stand or sit near others. You read it directly off their bodies.

Microbiome of the Body

The microbiome of the body may also play a role in our emotional life. A microbiome refers to various kinds of microorganisms that reside in the body including *pathogenic* kinds that cause disease in a host animal, *symbiotic* ones that are beneficial to both a host animal and the microorganism, and *commensal* ones in which the host animal or the microorganism derive some benefit. In particular, the gut microbiome in humans may affect one's emotions as well and interact with the information from the senses that is collated and processed by the brain. The gut microbiome refers to the microbes living in the intestines of humans and other animals that influences the brain through the endocrine and immune systems. It is estimated that there are 500–1000 bacteria alone that live inside the human gut along with fungi, viruses, and prokaryotes, single-cell microorganisms that lack a cell nucleus. However, over 10,000 microorganisms live inside the human body. The major axes of the brain-gut-body are made up of a number of cardinal systems including the autonomic nervous system (ANS), the central nervous system (CNS), the neuroendocrine and neuroimmune systems, the enteric nervous system of the gut, the HPA axis (hypothalamus, pituitary, and adrenal glands), and, of course, the microbiome of the gut. The gut microbiome may play a substantive role on brain development during pregnancy and in the postnatal period. Research suggests that the lack of breast-feeding in infancy and frequency of cesarean deliveries (about one-third of all birthings) may distort the infant's gut microbiome and produce subsequent health problems such as allergies, asthma, celiac disease, and diabetes, as well as obesity in older children and adults. Thus, the gut microbiome may be critical to both normal and abnormal development, particularly because it is intimately connected to other individuals in larger communities. Scientific studies in humans are attempting to understand the relationship between the brain and the gut and their role in child and adult disease states such as autistic spectrum disorders, chronic pain, anxiety spectrum disorders including depression, as well as metabolic disorders. For instance, it has been demonstrated that transcranial magnetic stimulation (dTMS)—applying a magnetic coil to the scalp—stimulates the brain and may actually improve the microenvironment of gut bacteria. It is being evaluated to treat obesity and other metabolic disorders. What is this connection between this mind/brain and the body?

We'll examine it next through the lens of the acquisition of physical skill.

Note

1. The quote, "The dynamic organization within ..." is from G. Allport, *Pattern and Growth in Personality*, 1961.

Bibliography

Allport, G. (1961). *Pattern and growth in personality*. New York, NY: Harcourt.

Andreasen, N. C., Paradiso, S., & O'Leary, D. S. (1998). "Cognitive dysmetria" as an integrative theory of schizophrenia: A dysfunction in cortical-subcortical-cerebellar circuitry? *Schizophrenia Bulletin, 24*(2), 203–218.

Berlucchi, G., & Aglioti, S. (1997). The body in the brain: Neural bases of corporeal awareness. *Trends in Neurosciences, 20*(12), 560–564.

Cole, J., & Waterman, I. (1995). *Pride and a daily marathon*. Cambridge, MA: MIT Press.

Damasio, A. R. (1996). The somatic marker hypothesis and the possible functions of the prefrontal cortex. *Philosophical Transactions of the Royal Society of London* (Series B), *351*(1346): 1413–1420.

Damasio, A. R. (2003). *Looking for Spinoza: Joy, sorrow, and the feeling brain*. New York, NY: Harcourt.

Daruna, J. H. (2012). *Introduction to psychoneuroimmunology* (2nd ed.). New York, NY: Elsevier.

Gallagher, S. (1986). Body image and body schema: A conceptual clarification. *Journal of Mind and Behavior, 7*(4), 541–554.

Goodman, N. (1976). *Languages of art: An approach to a theory of symbols* (2nd ed.). Indianapolis: Hackett.

Griffin, N. W., Ahern, P. P., Cheng, J., Heath, A. C., Ilkayeva, O., Newgard, C. B., Fontana, L., & Gordon, J. I. (2017). Prior dietary practices and connections to a human gut microbial metacommunity alter responses to diet interventions. *Cell Host and Microbe, 21*, 84–96.

Hein, G., & Knight, R. T. (2008). Superior temporal sulcus—It's my area: or is it? *Journal of Cognitive Neuroscience*, 20(12), 2125–2136.

Kolb, B., & Whishaw, I. Q. (2015). *Fundamentals of human neuropsychology* (7th ed.). New York, NY: Worth Publishers.

Linehan, M. M. (1993). *Cognitive-behavioral treatment of borderline personality disorder*. New York, NY: Guilford Press.

Lloyd-Price, J., Mahurkar, A., Rahnavard, G., Crabtree, J., Orvis, J., Hall, A. B., Huttenhower, C. (2017). Strains, functions and dynamics in the expanded Human Microbiome Project. *Nature*, 550(7674), 61–66.

Mischel, W. (2003). *Introduction to personality*. New York, NY: Wiley.

Noles, S. W., Cash, T. F., & Winstead, B. A. (1985). Body image, physical attractiveness, and depression. *Journal of Consulting and Clinical Psychology*, 53(1), 88–94.

Posner, M. I. (Ed.). (2012). *Cognitive neuroscience of attention* (2nd ed.). New York, NY: The Guilford Press.

Schmahmann, J. D., & Sherman, J. C. (1998). The cerebellar cognitive affective syndrome. *Brain, 121*(Pt 4), 561–579.

· 9 ·

PHYSICAL SKILL

Mental Practice Versus Physical Practice

Mental practice of athletic skills has substantial benefits over just physical practice alone. There are over 600 muscle groups in the body and potentially 2 × 10 × 600 possible combinations, which deeply impacts control of the body by physical practice alone. But in mental practice you have the ability to control both performance outcomes as well as environmental features of the specific sport with which you are engaged. You can slow down and examine your technique and skill in order to rectify performance issues and hone abilities. Meta-analyses of performance studies (in this case, a collection of 21 independent investigations) that have looked at the benefits of mental practice over the lack of mental practice, show an *effect size* of 0.68, which is fairly substantial indicating that mental practice is clearly efficacious. How is the effect size calculated? By adding up all the *mean* or average differences of the experimental groups (those that engaged in mental practice) and subtracting that from all of the mean differences of control groups (those that had no mental practice) divided by the standard deviation for all the groups, we arrive at the effect size. The standard deviation is simply the spread of scores around the mean (or average score) and is a marker of the amount of dispersion. As it turns out, though, use of internal kinesthetic imagery by athletes is actually more effective in heightening performance than external imagery. If a swimmer, for example, kinesthetically represents to herself the goal at the other end of the pool of changing direction underwater and pushing off the far side of the pool and imagines the feel and experience of being in the water, then she is more likely to have better performance. If instead, she imagines watching herself perform the laps in the pool, her performance improves but not as much as if she had used kinesthetic imagery.

Why is there a difference? When using kinesthetic imagery, the relevant muscles are activated and learn to perform the same skill as those pictured in the mind based on retrieving appropriate musculature information in kinesthetic memory. Edmund Jacobson, an American internist, demonstrated in 1931 that when one imagines bending an arm or raising a leg electromyographic (EMG) activity is produced in the relevant musculature. Then, afferent fibers from the sensorium (the five external sensory modalities) provide sensory feedback that inform the temporal cortices where kinesthetic memory is activated. This provides a mechanism for response preparation and readies the musculature for actual movement. At the same time, when engaging kinesthetic imagery, you are also acquiring spatial, temporal, and sequential aspects of the movement that assist in learning the symbolic aspects of forming and maintaining the kinesthetic imagery in one's mind. During mental practice you are importing the overt behavior into a collection of kinesthetic images for use in acquiring the specific skill in a particular sport. All of these embodied aspects of learning a skill are largely inactivated when using just external imagery. But, when using kinesthetic imagery, a bodily skill is thus embodied, by definition.

Mental practice is used both in the acquisition as well as the performance of the actual physical skill. In the performance of a well-learned physical skill there is a preparatory phase in anticipation of the use of the skill and a rehearsal phase that abets its successful retrieval from memory as well as reinforces the successful completion of the movement. In the preparatory phase, the MBB makes use of kinesthetic imagery, attends to specific proprioceptive sensations in the musculature, and ramps up preparatory arousal for maximal performance. In the rehearsal phase, the MBB ramps up its cognitive and rehearsal strategies in order to fully realize its goal. These cognitive strategies include correction of errors and mental consolidation of strategies for the greatest performance gains, which might include, for example, changing one's golf swing, repositioning the feet, reshaping the way the club is held with the hands, re-balancing the body, and the like.

Of course, mental practice is no substitute on its own for continuous physical practice, but a number of logical consequences follow. First, individuals with a greater ability to image movement tend to have better performance when using mental practice. Second, mental practice is highly effective in the acquisition, retention, and in the performance of learned physical actions, activities, and movements. Third, there are general strategies for both learning motor skills as well as transferring them to related sets of skills and mental practice is an important aspect at every stage. These include preparatory strategies, use of visual and kinesthetic imagery, learning to concentrate on relevant physical cues, actually executing the action, and evaluating knowledge of results that includes correction of errors and the updating of cognitive strategies so as to augment later performance.

A contrarian view is runner's cognitive strategies for peak performance. Long-distance runners, in particular, tend to focus on external stimuli during their training experiences thus cutting themselves off from environmental cues, external imagery, and feedback from the body. But, as it turns out, it serves as a way of sustaining their repetitious gait over long distances as well as a way of bracketing considerable bodily pain. To be sure, I wasn't just thinking about 26.2-mile marathon runners here, which is painful enough, but ultra-marathoners that run 100 miles in the desert or continuously for 24 hours.

In a similar vein, surgical residents with higher scores on form board and mental rotation tests—measures of visuospatial abilities—perform better on surgical tasks and are able to transfer their surgical skills to more complex surgical functions. For one, both tests are tests of mental imagery, and surgeons who excel on these tests, tend to be better surgeons. Second, the two tests involve moving actual objects—or visuospatial memory of working with similar three-dimensional forms in the past as familiarity with these stacked cubes on the mental rotation tests radically increases success—as well as mental objects in three-dimensional space as that reciprocity and feedback between mind and body is crucial. In essence, you are mentally simulating action and, one might say, enactively simulating mentation.

In all of these mental-physical tasks, both feedforward and feedback mechanisms are required because afferent (sensory) feedback cannot alone guide the initial phase of movement but necessitates feedforward models that map efferent (motor) output onto its sensory consequences. Yet, it also requires inverse models (inverse feedback) that map sensory results back onto their motor output. It's reciprocal or, that is to say, the integration of the efferent output with reafferent sensory input. Or to put it another way, it enables us to distinguish between sensory input from the environment—known as *exafference*—from sensory input from our own self-generated actions, or *reafference*, resulting in an efference copy that helps us tell the difference between the two, which I previously discussed.

What we haven't looked as yet is body therapies, which we will begin to dissect in the following chapter.

Bibliography

Craig, C. M., Delay, D., Grealy, M. A., & Lee, D. N. (2000). Guiding the swing in golf putting. *Nature*, *405*, 295–296.

Goleman, D. (1994, October 11). Peak performance: Why records fall. *The New York Times*, C1, C12.

Hinshaw, K. E. (1992). The effects of mental practice on motor skills performance: Critical evaluation and meta-analysis. *Imagination, Cognition, and Personality*, *11*(1), 3–35.

Jacobson, E. (1938). *Progressive relaxation*. Chicago, IL: University of Chicago Press.

Magill, R. A. (1989). *Motor learning: Concepts and applications* (3rd ed.). Dubuque, IA: William C. Brown.

Ogles, B. M., Lynn, S. J., Masters, K. S., Hoefel, T. D., & Marsden, K. (1994). Runner's cognitive strategies and motivations: Absorption, fantasy style, and dissociative experiences. *Imagination, Cognition, and Personality*, *13*(2), 163–174.

Ratey, J. J. (2008). *Spark: The revolutionary new science of exercise and the brain*. New York, NY: Little, Brown.

Wanzel, K. R., Hamstra, S. J., Anastakis, D. J., Matsumoto, E. D., & Cusimano, M. D. (2002). Effect of visual-spatial ability on learning of spatially-complex surgical skills. *The Lancet*, 359(9302), 230–231.

Willingham, D. B. (1998). A neuropsychological theory of motor skill learning. *Psychological Review*, *105*(3), 558–584.

· 1 0 ·

PEDAGOGY AND BODY THERAPIES

The role of movement, action, and activity in learning and pedagogy and the role of the mind embodied in sport and body therapies, like the Pilates, Feldenkrais, and Alexander methods as well as Yoga and other Eastern practices, are taken up in this chapter. These skills and practices are important to the lives and hearts of tens of millions, so they deserve a reasoned examination.

Learning and Pedagogy

Let's start by looking at the way individuals learn. We can visualize one type of learning as a *S-shaped curve*. In the initial stages of learning a cognitive or physical skill there appears to be little skill acquisition. Learning the tango is a good example. It's a difficult dance and involves coordinating rapid steps with a partner along with accompanying arm, limb, and trunk movements. However, once the subskills begin to be mastered and coordinated, there is a brisk update of whole-body movements as one masters the dance but then a slowing down again as the skill plateaus. Another individual learning the tango, however, might experience a sudden acceleration of learning after a period of relatively little progress as is often true in complex skill acquisition. This is known as a *salutatory learning curve*. Other dance enthusiasts might experience a fluctuating period of seemingly moving forward in learning some of the complex skills of Argentine tango while other subskills such as keeping in sync with one's partner may lag behind. This is referred to as a *stepwise learning curve*. A *classic learning curve* assumes, of course, that someone just gets better and better with time in learning a skill, whether cognitive or physical.

In all of these instances, there is both an underlying latent capacity of previously acquired skills and subskills—latent abilities that are hidden and unobservable—as well as a surface layer of manifest abilities that reflect emerging patterns in development as well as specific responses to instruction and skill acquisition. So, one way to think about abilities and skills and their acquisition is to conceive of them as *forms of developing expertise*. They are flexible and malleable. They imply latent capacities that are emerging—not yet developed abilities—but developing skills. They strongly suggest that the potential for learning itself changes at different stages of acquisition and is therefore continually dynamic in its growth and elaboration. Instruction becomes especially critical because how the skill is learned and scaffolded through engagement with an instructor or teacher determines the trajectory of its growth. Difficulties in learning a particular skill or subskill, for instance, may be due to limited working memory in that particular skill area and not an indicator of a skill deficit. Personality factors, cultural background, and metacognitive abilities—the ability to think about and question what one is learning—may also abet or attenuate emerging skill acquisition and use. Keep these ideas in mind as we examine below various kinds of physical and bodily skills.

One of the most effective ways to relieve stress, irrational thoughts, and negative feelings as well as alleviate many kinds of medical conditions is through so-called body therapies. Sports and dance are the most common examples, but there are others. Instructed therapies such as the Alexander technique, the Feldenkrais method of body awareness, and Pilates all work the body systematically through the MBB. So, does the many varieties of therapeutic massage and Yoga as well as other Eastern practices such as Tai chi and buteyko or the use of breathing exercises. Just about all of these methods are taught therapies and involve the use of various kinds of instruction realized through movement, action, and activity.

On the other hand, stress and a sedentary lifestyle, poor diet and nutrition, excessive weight, and exposure to pathogens and toxicants (alcohol, tobacco) are major risk factors for a number of medical conditions including cardiovascular disease (CVD), cancer, diabetes and metabolic disorders, mental and brain health, arthritis and musculoskeletal conditions, respiratory conditions, attentional issues, insomnia, and smoking cessation. Do body therapies have a positive effect on these conditions? All of the therapies I will discuss below are physical or *bodily therapies* that involve movement of the whole-body, trunk, limbs, face or breath.

Yoga[1]

The Indian tradition, of which Yoga is a part, dates back to 5000 B.C.E. and is rooted in Ayurvedic medicine or "knowledge of longevity." Ayurvedic medicine emphasizes mental, physical, social, and spiritual harmony. One's state of mind is made up of a balance of three mental principles: Sattva (the mind's natural state), tamas (a focus on external objects), and rajas (a focus on action). A predominance of tamas and rajas suppresses sattva and causes the mind to be more "efficient" improving mental health. The body's dosas (vata, pitta, and kapha), however, work together to sustain physical health and are representative of the five central elements of both the universe and the human body (earth, water, fire, air, and space). Vata directs action, pitta directs transformations, and kapha stabilizes the two. The balance or imbalances of the dosas maintain physical health or create disease. Treatment involves stimulation of the five senses (music, massage, and the use of aromas), purification (enemas and emesis or vomiting), and the use of medicines, which can be either actual physical substances or *certain thoughts or beliefs* all aimed at balancing the dosas.

Yoga is a mind-body practice that combines breathing techniques, physical postures or asanas, and meditation to re-establish mind-body balance and improve overall health. It has eight dimensions including *pranayama* or breathing, *asanas* or postures, and *dhyana* or meditation. Yoga has many styles of practice including Hatha yoga, Bikram or hot yoga, Vinyasa, and Ashtanga yoga. These various styles focus on different elements of breathing, physical postures, and meditation as well as the amount of intensity of the exercise.

Ashtanga yoga is an intense and difficult yoga to learn and often devotees take 4–5 classes a week. It is often referred to as power yoga. The role of the instructor is to guide movement as well as assist in correcting the postures, limb and trunk alignments and adjustments, and breathing. The flow of the sequence of postures is referred to as *vinyasa* and is another central aspect of training. It involves a set of subskills including smoothly executing individual postures, proper limb and trunk alignment, and appropriate breathing. The latter typically involves inhaling for 10–15 seconds and then exhaling for

10–15 seconds, but in unison with the posture and in flow with the sequence of asanas. These are the major skill sets. The complexity in the practice of yoga probably rivals learning the tango. To get a better understanding of physical complexity, let's look for a moment at the development of dance movement in preschool and school-age children.

My own research on aesthetic movement has classified the various ways that children express thoughts and ideas through dance. At a general level, the research indicates that 3- and 4-year-olds lack the ability to express tension or weight in movement. The ability to express through movement the concept of tension presents a preschooler with a number of problems in using the body to exemplify straining, tautness or rigidity although they can roughly imitate the action when demonstrated to them by an older child or adult. Likewise, the exemplification of weight is a fairly complex enactive problem for them to figure out unless it has been previously shown to them. However, in terms of motor organization, they have developed front-back but poor lateral movement; have acquired quasi-skipping, -marching, and -jumping abilities; demonstrate both asymmetric use of the body and body parts but lack more advanced balancing abilities. In terms of cognitive capacities, they can treat one movement as another, represent a movement for an absent concept, and distort a property of a movement so as to treat it as another movement property. By 5- or 6-years of age, however, children have acquired lateral movement, complex movements such as skipping, can coordinate movement with and around objects, use their upper body to propel themselves forward in a horizontal plane, and demonstrate the first and second positions in modern ballet technique. In addition, they can create geometric shapes with body parts, demonstrate fairly complex metrical properties of rhythm in their movement, show symmetric use of the body and body parts, and have acquired more advanced balancing abilities. In both age groups, studies of nonverbal abilities have found that children will make use of a body part to stand in place of a concept or idea—or what I called earlier metaphoric gestures—use simple deictic and spatial gestures, depict in movement a bodily action or *kinetograph*, and use gestures that regulate action between group participants or what are known as elementary *regulators*.

Like dance, yoga involves complex movement skills and is thought to help in stress reduction and it may even be an effective strategy in the prevention of cardiovascular disease, among other disorders. Both the *Cochrane Review* and the *Australian Government Department of Health Review of Natural Therapies* were consulted, and the results appear below. Cochrane is a non-profit, non-governmental organization (NGO) that compiles evidence-based findings in many areas of medical research. In 2015, the Australian Government did a major survey of natural therapies focusing on clinical efficacy, safety, quality, and cost effectiveness. The studies were examined by experts from independent organizations who did systematic reviews of the literature and made available their findings to the public.

Cochrane Review

Yoga had positive effects on diastolic blood pressure (the part of the cardiac cycle when the heart refills with blood), HDL or good cholesterol, triglycerides (markers of atherosclerosis), but unclear effects on LDL or bad cholesterol. There is some evidence that yoga improves back-related function after three to six months of regular practice and may be slightly effective for low back pain. Mind-body interventions including yoga and hypnotherapy may have some benefits for the management of anxiety during pregnancy. There was no evidence, however, that meditation-based activities like yoga are helpful with disorders such as ADHD that involve chronic levels of hyperactivity, inattention or impulsiveness. Nonetheless, exercise therapy, including yoga, may reduce fatigue in patients with multiple sclerosis.

Australian Government Department of Health Review of Natural Therapies

There was weak evidence that yoga has positive benefits for individuals diagnosed with depressive disorders. There was insufficient evidence to draw any definite conclusions for other clinical conditions including arthritis and musculoskeletal conditions, cancer, cardiovascular disease, insomnia, menopause, mental health conditions, neurological conditions, ADHD, pregnancy and labor, renal disease, respiratory conditions, smoking cessation, and type 2 diabetes and metabolic syndrome.

Alexander Technique

The Alexander technique is a form of mind-body therapy that is designed to correct poor posture and bring the body into appropriate alignment through a succession of movements to help it function more effectively and reduce pain.

Through individualized assessment and a series of progressive, calibrated lessons, the patient releases undesired tension in the body (head, neck, and spine) and begins to improve postural tone and neuromuscular coordination. Lessons are made up of hand contact by the instructor, as well as verbally-guided movement while standing, lifting, walking, sitting, and lying down. It reportedly produces relaxation and may help individuals with asthma breathe more easily.

Cochrane Review

The Alexander technique has been prescribed to individuals with asthmatic symptoms that are caused by air pollution, allergies, pollen, and stress. The technique purportedly improves breathing because during an asthma attack the bronchial airways become constricted resulting in breathing issues, such as coughing and wheezing. However, there was a paucity of evidence demonstrating the efficacy of the Alexander technique over traditional medication (inhalers) for asthma to reduce symptoms.

Australian Government Department of Health Review of Natural Therapies

The Alexander technique may be effective in alleviating disability and pain in individuals with low back pain. This effect may last for up to three months, but the longer-term effectiveness is unknown. However, improvements that are statistically significant have been reported for the intensity and length of pain measured in median number of days, extent of disability, and persistent positive mood when compared to nursing care, self-help groups, and physical exercise. For other clinical conditions such as neck pain, chronic musculoskeletal pain, asthma, post-traumatic stress disorder, and generalized anxiety disorder there was no conclusive evidence.

Feldenkrais Method

The Feldenkrais method attempts to improve posture, breathing, and movement by combining gentle touch with hands-on verbal instruction. Individuals are taught to focus on their breathing and movement with the aim of bringing mindfulness to these activities and thus adopting new habits. Through these practices, individuals are said to experience an improvement in their overall sense of well-being. There are two formats for teaching the Feldenkrais method. *Functional Integration* requires one-on-one, hands-on sessions during which the practitioner gently moves the patient into various positions. *Awareness Through Movement*, on the other hand, teaches individuals using recorded instructions either individually or in a group setting. Pedagogical instruction is often conducted standing, sitting in chairs or lying on a carpet.

Cochrane Review

No extant studies available.

Australian Government Department of Health Review of Natural Therapies

Only studies of women with work-related complaints of the neck and shoulder, people with chronic low back pain, and older individuals at risk of falling have been reported in the literature. Randomized clinical trials reported a positive effect favoring Feldenkrais compared to no treatment, sham, or one's usual activities. The studies, however, were small and statistically underpowered and the level of confidence in the evidence was low. There is also a paucity of evidence regarding the effectiveness of Feldenkrais for the improvement of health outcomes for other clinical conditions. The Feldenkrais method is intended to re-train the brain to improve posture and movement with the goal of improving quality of life. The effectiveness of Feldenkrais for improving health outcomes in people with chronic musculoskeletal pain (neck, shoulder, or knee), fibromyalgia, motor skills in patients after stroke, post-traumatic stress disorder, and adults with generalized anxiety disorder, is unknown.

Pilates

Joseph Pilates (1883–1967) was a physical trainer of German origin who came to believe by his mid-teens that posture, breathing, and lifestyle components were key to mental and physical health. After arriving in the US in 1925, he set up shop in New York City and focused on his work with students including many modern dancers and ballerinas. In his mind-body practice, he emphasized the trunk postural muscles of the core as highly important for supporting and aligning the spine and the key to overall balance of the body. He taught breath awareness and accentuated the primacy of the abdominal and deep torso musculature. His method comprises about 500 exercises using specially constructed equipment and floor mats. Exercises focus on centering and strengthening the core musculature, control of postural integrity, precision in application of the exercises, and mental and physical concentration. He also placed great emphasis on isometric resistance in which there is contraction of the musculature without moving the joints. This is similar to executing a Yoga plank (posture) that emphasizes muscles in the core or holding a weight in a fixed position over time. There is also significance given to stabilizing muscles that improve posture and condition the body. The putative benefits of the Pilates method include enhancement of coordination and strength, flexibility and balance, symmetry and definition of the musculature of the body, increased range of motion, and improvements in one's proprioceptive sense. Thus, one might assume that the Pilates method would be of therapeutic benefit to individuals with musculoskeletal conditions or even those individuals with Parkinson's disease affecting balance.

Cochrane Review

The Pilates method is somewhat effective for low back pain and disability in the short and intermediate term as well as demonstrating improvement in overall functioning. However, it is probably no more effective than other forms of physical exercise for pain and disability, which as it turns out, may in some cases, be more effective in both the short and intermediate term.

Australian Government Department of Health Review of Natural Therapies

The Pilates method may help some adults with low back pain and improve overall health outcomes. But, for other clinical conditions such as being overweight or obese, breast cancer, falls prevention in older adults, stress, and urinary incontinence in women, there is little evidence to support its use given the dearth and quality of currently available studies.

Tai Chi

The Chinese tradition of healing dates back to 4500 B.C.E. and is rooted in the Taoist philosophy that the body is a miniaturized version of the universe.

While change is common to both, it is represented in the body by a flow of energy called "qi." Lack of *qi* makes one susceptible to illness because of constantly evolving circumstances that make one unable to deal with either changes in the environment (an argument with a spouse) or changes within the self (angry feelings). The balance of energetic forces distributed throughout the body is represented by the polar opposites of ying and yang. If these energetic forces are unbalanced, it is because *qi* has been blocked and treatment consists of redistributing *qi* throughout the body. To restore mental and physical balance, treatment consists of nutritional changes to one's diet, exercise such as Tai chi, as well as adopting a correct psychological attitude.

Tai Chi or *Taijiquan* as it is known, is a mind-body practice that originated in China and encompasses relaxation, deep breathing, and slow and gentle physical movement. There are many different styles of Tai Chi, but each is based on the same underlying principles. In addition to movement of the body in various postures, the practice of Tai Chi also focuses on mental concentration and deep breathing. Tai Chi is often practiced in groups as well as individually and outdoors. It has been used to treat musculoskeletal conditions such as rheumatoid arthritis.

Cochrane Review

There is evidence that Tai Chi has a number of benefits for the body particularly those with rheumatoid arthritis. It improves the range of motion of the knee, hip, and ankle. But it does not rehabilitate joint tenderness, grip strength, swollen joints or an individual's ability to perform daily chores. There was some evidence, however, for better functional capacity and pulmonary function in chronic obstructive pulmonary disease (COPD).

Australian Government Department of Health Review of Natural Therapies

There is evidence that Tai Chi has some positive health benefits for improving muscle strength in late-age adults as well as enhancing quality of life in individuals with cardiovascular disease, hypertension, and osteoarthritis including accompanying pain. On the other hand, there is little extant evidence to suggest that Tai Chi is effective for reducing Type 2 diabetes or in preventing falls in older adults.

Buteyko (Breathing Exercises)

The Buteyko breathing technique was invented by a Russian physician, Konstantin Pavlovich Buteyko (1923–2003), for the treatment of asthma and other respiratory conditions. The technique is an instructional process in which patients are initially coached in controlled shallow breathing through the nose followed by holding the breath at the end of the exhalation.

Not surprisingly, breathing exercises have been taught for various conditions in Western and Eastern cultures for hundreds if not thousands of years. Anecdotally, mind-body practices emphasizing breathing exercises suggest an alleviation of symptoms of asthma.

Cochrane Review

Six of eleven studies assessing lung function demonstrated a favorable difference for breathing exercises for asthmatic symptoms. Generally, no adverse effects were reported, and the treatment was deemed safe and well-tolerated.

Australian Government Department of Health Review of Natural Therapies

The reduction of bronchodilator use, in patients with asthmatic symptoms through use of the Buteyko breathing technique, is often accomplished in practice, but there is little or no effect on core asthma symptoms or pulmonary function. Thus, there is little support for use of the technique in managing the overall symptoms of asthma.

Therapeutic Massage

The Greek tradition of medicine is associated with the teachings of Hippocrates and his followers including Galen, a prominent Roman physician, starting around 400 B.C.E. Central to health was the balance of the four humors (blood, phlegm, yellow and black bile) and the balance of the four elements (air, water, fire, earth). Each humor had a different origin in the body (heart, head, gall bladder, spleen, respectively) and were renewed by food and water. Health and disease resulted from their imbalance. For instance, excessive black bile caused melancholia or depression and produced various forms of malignancy. Humors were generated, blended, and transformed by way of the body's internal fire or cauldron, which originated in the left ventricle of the heart. The humors gave rise to the four temperaments or what is now referred to as *character* or *personality* as I discussed in Chapter 8. These sanguine, phlegmatic, choleric, and melancholic temperaments, moreover, were believed to be affected by the seasons of the year. Treatment consisted of restoring the balance of the humors through a proper conduct of life that included appropriate food and drink, suitable daily activities and living conditions, proper exercise, the quality of one's sleep, engagement in healthy sexual activity, as well as psychological understanding of oneself and others.

Therapeutic massage as practiced in Western cultures involves manipulating or kneading the musculature of the body or other soft tissues for improving health and well-being. It is a form of *manual therapy* that involves applying pressure to the muscles, ligaments, tendons, and fascia or sheaths of fibrous tissue—generally collagen—that enclose, attach, and stabilize various muscle groups. There are many forms of therapeutic massage, but they all vary around the amount of pressure and touch as well as the intensity of the intervention.

Cochrane Review

Therapeutic massage had some benefits in alleviating low back pain and improving function in the short-term but not in the long-term. There was no evidence, however, for the alleviation of pressure ulcers caused by constant friction or pressure. However, massage for mechanical neck pain provided alleviation of both tenderness and pain in the short-term. This is significant because neck pain is quite common and massage—manipulating the soft tissues surrounding the neck with the elbow, arm, and hand—is generally effective.

Australian Government Department of Health Review of Natural Therapies

Therapeutic massage was effective in the short-term, but not long-term, for chronic low back pain and in short-term relief of pain in patients with acute low back pain. In preterm infants, it is effective for promoting weight gain and reducing the length of stay in the hospital.

To briefly summarize, all three ancient traditions (Ayurvedic, Chinese, and Greek) emphasize health as consisting of:

1. A balance of substances
2. The individual integrated into the larger social and physical world

3. Prevention as the goal of medical knowledge
4. The individual's ability to self-heal
5. Treatment conceptualized as treating the whole person

Unfortunately, there are limited studies of the effectiveness of natural body therapies for mind-body disorders including medical conditions and brain health. Almost all of these therapies involve extensive learning and practice to gain the health and exercise benefits. Much value would be accrued if future researchers looked more deeply and systematically at a broad range of natural body therapies.

Note

1. I took yoga classes in college after being introduced to it in a course on *Buddhism of China and Japan* by an Asian scholar before it became popular and a national obsession. A few years later, when I was a graduate student in NYC, I took classes in the Alexander technique, Feldenkrais method, and Tai Chi. Many years after that I took private lessons with a young Pilates instructor at an Equinox Fitness Club on the Upper West Side of Manhattan where I used to get a therapeutic massage each week after working out at the gym.

Bibliography

Carlson, S., & Seitz, J. A. (1998, August). *The development of kinesthetic intelligence in children.* Paper presented at the New York Academy of Sciences, New York, NY.

Cochrane Library accessible at: https://www.cochranelibrary.com.

Department of Health, Australian Government (2015). *Review of the Australian Government rebate on natural therapies for private health insurance.* Canberra, Australia: Department of Health.

Mirani C., & Seitz, J. A. (1998, August). *Development of bodily-gestural skills in children.* Paper presented at the New York Academy of Sciences, New York, NY.

Seitz, J. A. (1992). The development of bodily-kinesthetic intelligence in children: Implications for education and artistry. *Holistic Education Review*, 5(2), 35–39.

Seitz, J. A. (1994, August). *The development of aesthetic movement skills.* Paper presented at the meeting of the International Association for Empirical Aesthetics, Montreal, Canada.

Seitz, J. A. (1996). The development of aesthetic movement: Linkages to preschool education. *Journal of Early Education and Family Review*, 5(3), 7–9.

Seitz, J. A. (2002). Mind, dance, and pedagogy. *Journal of Aesthetic Education*, 36(4), 37–42.

Seitz, J. A. (2018). Neuroscience and brain and mental health blog accessible at: https://www.linkedin.com/in/dr-jay-seitz-b6b4408/detail/recent-activity/posts/

Sternberg, R. J., & Grigorenko, E. (2001). *Dynamic testing: The nature and measurement of learning potential.* NY: Cambridge University Press.

· 11 ·

THEORETICAL DISCUSSION

Evolutionary Origins of Complex Cognitive Abilities

What if music, aesthetic and athletic movement, image-making, speech and language, sociality, and number had a common origin? All six are complex cognitive abilities and why should one be more equal than the others, to paraphrase George Orwell?[1] We don't just think in words but in musical and non-musical sounds; motor impressions; visual, gustatory, and haptic images; social relations and self-examination; and numbers and their operations. We know this from introspection, from everyday experience including socializing with friends, discussing wines with an oenophile, hearing a lecture from a docent at an art museum, attending the ballet, listening to music on your smartphone, from case studies of artists and scientists, among others, and from diverse experimental, medical, and scientific evidence. We know that language is constituted, not just by words, but by rhythm and timing (meter), intonation and stress (accent) on words (melody), changes in loudness and sound intensity, facial displays, and gesture and posture, that is, the paralinguistic cues and vocal prosody of everyday speech.

One likely possibility in human evolution was that there was a primordial matrix of *protospeech, protonumber, protoart, protosocial, protomusical, and protoaesthetic movement forms* in early hominids. There are remnants of these capacities in other species including the African Grey parrot (speech), rhesus monkey (number), elephant (visual art), meerkat (sociality), whale (music), bonobo (logic and number), and birds and canines (dance; if you don't believe me, Google the "Dancing Cockatoo" or "Kasperlicious, the Dancing Poodle") that undergird modern human intelligence. These intellective or *noetic forms* co-evolved, more or less together, resulting in modern human intellect; speech and language; numerical competence including the capacity for quantitative navigation; visual and spatial intelligence; social abilities including emotional, social and Machiavellian intelligence; and athletic and aesthetic movement. Their co-evolving together was responsible for the efflorescence of complex cognitive abilities in modern hominins in the last 500,000 years or so. Scientific and medical studies, and theories, that examined each of them independently from each other have, created an illusion in evolutionary theory. Like the human brain and body, it's the connections that matter. And the coevolving of each of these noetic forms, correlated and together, is responsible for the perception that these noetic forms exploded in the late Pleistocene period. They worked synergistically in creating the complexity, architecture, and processing capabilities of the modern hominin brain. These initially relatively encapsulated cognitive streams in vertebrate and invertebrate evolution also facilitated the emergence of intermodular or metaphoric (creative or *supramodular*) thought and the cognitive flexibility found in the mind-brain-body (MBB) of modern hominins.

The emergence of *image-making* in early humans may have been the key ingredient that allowed the generation of geometric forms (graphemes, shapes) in picture-making (prehistoric cave art) and habitat construction (formally, architecture), and eventually written language. Number builds on the basic rhythmicity of the mammalian MBB that revolves around circadian and disparate biorhythms (the pulse or beat of life). Music as well as both aesthetic and non-aesthetic movement binds human groups rhythmically as well as provides a vehicle for individuals to experience as well as express sundry emotional states. Non-aesthetic movement (collectively, athletic movement and sports) not only bonds human groups but creates situations for social competition and cooperation; displays of strength and endurance; inter-individual ludic, physical, and amorous interaction; and the like.

The evolution of different forms of memory to support these higher-order cognitive abilities would have included a significant number of diverse mnemonic forms. Semantic and episodic memory, that is, memory for facts and memory for events, prospective memory or remembering to do something in the future, social and emotional memory, numerical memory, visuospatial memory, kinesthetic or muscle memory, and musical memory. These would have predated the evolution of language and other complex cognitive abilities as they appear in non-human animals (whales, corvids, canines, and rats).

Structural Characteristics of the Neocortex

The modern human adult brain contains, on average, about 86 billion neurons of which approximately 16.34 billion (19%) make up the grey and white matter of the cortex (Figure 11.1). The current estimates are there are about 6 billion+ neurons in the gray matter of the cortex and 10.3 billion neurons in the white matter with supporting neuroglia in both (8.7 billion and 20 billion, respectively). In case you're wondering, the current estimates are that there are about 100 trillion (10^{14}) synaptic (cell-to-cell) connections in the human adult brain. It's not quite, but almost all about the connections and well over 1100 times more.

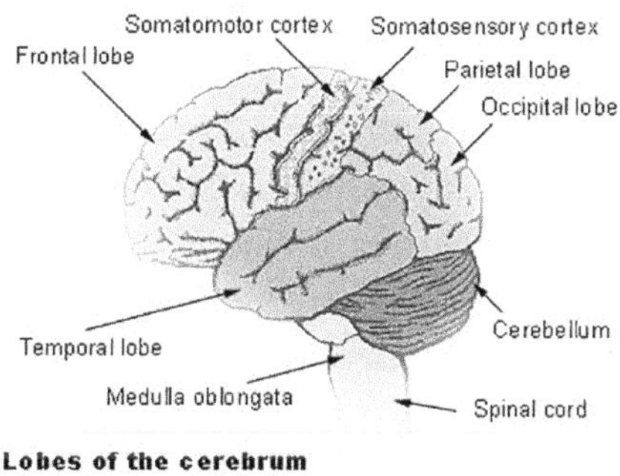

Figure 11.1: Neuronal Count in the Human Brain.
Credit: https://commons.wikimedia.org/wiki/File:Illu_cerebrum_lobes.jpg. Public domain.

Robin Dunbar, a British anthropologist, first proposed in the 1990s that there was a significant correlation between average social group size and average primate brain size. It was subsequently referred to as Dunbar's number. By computing the average human brain size—approximately 1260 cubic centimeters in size and 1500 grams in weight—and extrapolating from primate brain and group sizes, he suggested that humans can manage about 150 relatively enduring relationships with others in which you know something about each individual as well as something about how each individual relates to each other.

This is so because the modern human brain is severely limited in what we can store in memory. By middle-age we probably remember about 1/10,000 of a lived life amounting to about 215,000+ seconds (60 hours) where each second roughly represents a single memory. Vocabularies average about 50,000 words along with chunks of oral texts, episodes are made up of objects and actions, images are largely primal sketches, music is familiar tunes stored as large musical chunks, numerical awareness is quite small even though we can count up large numbers, fine and gross movements are stored as motor plans, and our social world is made up of about 150 people or less who we know in varying degrees.

About 2 1/2 days of disparate information. And, this information store is constantly changing and updating itself. The belief that it is much, much larger is based on the fact that we lived in a networked world of people, technologies, and sources of external information constantly accessed through the MBB that allow us to feel a sense of a much larger world. But, live out on the desert or in the woods for a month or more without any external memory aids or technology and your world will quickly shrink to its seemingly truer size.

As for the white matter, it comprises the parallel, distributed connection system of the grey matter or cerebral cortex (cortical mantle). *Most white matter fibers are relatively short* in length and connect neighboring grey matter cell bodies. About 2%, however, are *long association fibers* that connect neighboring brain regions as well as connect the two halves of the brain, also known as *callosal or commissural fibers*. The last group are *projection fibers* that connect with the spinal cord and lower brain regions including subcortical areas. These association, commissural, and projection fibers are often referred to as *white matter tracts*. Tracts can be further subdivided into commissures such as the anterior commissure connecting the two halves of the temporal lobe located in the left and right halves of the brain and the hippocampal commissure connecting the two halves of the hippocampus. Another type of tract is so named a *decussation or crossed-fiber tract* such as the corticospinal or corticobulbar tracts that cross deep in the brain at the level of the medulla oblongata, an area known at the medullary pyramids. As a general rule, the number of fibers of a particular length is inversely proportional to their length. Or, the longer the fibers the fewer there are, which is saying the same thing if you're reading closely.

About 69 billion or so neurons are in the cerebellum. There are another 85 billion or so neuroglial cells in the whole brain that provide support for these neurons. *Neuroglial or glial cells*, of which there are four main classes, have extremely important functions. They provide structural support, nutrients, and oxygen to neighboring neurons and guide early brain development. During neural transmission, they clear neurotransmitters from the synaptic cleft or gap that lies between one neuron and the next and very recent research has suggested that they modulate neural transmission by secreting their own gliotransmitters such as glutamate and ATP. It appears that the overall cell composition of human primate brains is no different from other primates and that "the human brain is an isometrically scaled-up primate brain."

Even the number of neurons in the human brain is not linearly related to increasing size. Size and number of neurons matter up to a point, but it's *modularization* and the *connectome* (the overall connections) that really make a difference along with the *cytoarchitecture* (structural characteristics) of disparate brain regions. Human brains are highly interconnected but the individual brain regions change in processor function from one area to the next. Now imagine a computer with not only many different computer processors (multicore and multithreaded) but many different *kinds* of processors depending on what kinds of computing functions they were required to do. Individually programmed tasks of the same subroutine may be processed in different cores of the computer and that is called *multiprocessing*. Using just a single core using subunits or threads in the core it's termed *multithreading*. If tasks are processed concurrently it is called *parallel processing*. We're just beginning to see this in consumer electronic products where a computer may have various co-processors along with the main computer processor. That's the human primate brain. It's a multiprocessor, multithreaded neurocomputer.

The cerebral cortex itself is about 2500 cubic centimeters (cu^3) in volume and the neocortex, the last evolving part of the cerebral cortex (the forebrain), is about 1800 cu^3 and 2–5 mm in thickness and possesses six laminated layers (I–VI). Layers I–III are known as the supragranular layers and represent the origin and termination of connections within the cortex or what are called *intracortical connections*. Some are *associational* (connect cortical sites in the same hemisphere) and some are *commissural* (connect cortical sites in the opposite hemisphere). Layer IV receives connections from the thalamus, which relays information to and from the cortex and hence are referred to as *thalamocortical connections*. Layers V–VI primarily connect the motor cortices or agranular cortex to subcortical regions including *efferent projections* to the spinal cord, brain stem and basal ganglia, as well as to the thalamus. The neocortex (or *neopallium*) is not as thick or as large as you might imagine but it comprises the most advanced areas of the human neurocomputer for storing and processing information. The rest of the cerebral cortex is comprised of *allocortex* including the hippocampal complex for encoding memories and made up of four laminar layers known as *archicortex* and the oldest part of the cortex. The olfactory system is made up of three laminar layers but missing the granular layers II and IV of the neocortex, and known as *paleocortex*, somewhere in evolutionary age between the older archicortex and the newer neocortex.

Algorithms, Recursion, Pattern Recognition, and a Rule-Based Logical System

Each of these *noetic forms* or complex cognitive abilities have *algorithms or rules* for combining and sequencing elements as well as the capacity for recursion. In the first instance, there are rules of syntax (*Merge*), mathematical and navigational rules, rules of visual thinking (line, form, color; see, for instance, Josef Albers' *Interaction of Color*, 1963), rules of social interaction, tonal knowledge rules, and neuromuscular rules as I described in Chapter 1. In the second instance, there is recursion in number ($A + (B + C) \times D^2$), in language (words embedded in phrases embedded in sentences), in social interaction or *theory of mind* ("What is Mary thinking about what Tom is thinking about Sue?"), music (a musical motif embedded in a larger musical segment; Richard Wagner's *Ring Cycle* is loaded with 60 leitmotifs; check out the score online), navigation (dead reckoning or path integration, route knowledge, survey knowledge, and landmarks), motor organization in movement as described in Chapter 1, and image-making (lines embedded in forms embedded in color regions).

All of these noetic forms or systems are hierarchical, combinatorial/componential, and recursive. That is, sequential, hierarchical processing of multiple forms of information. Is there a common *code* (algorithm, rule) or underlying universal, rule-based logical system as Ray Kurzweil, the computer scientist, suggests, "plasticity implies a common algorithm across the entire cortex" or are there many different but overlapping codes in the MBB? Or, is the capacity for recursion the common algorithm that lies across the entire cortex because structurally—that is, smaller units within larger ones that repeat the same procedures—the cortex itself is recursive?

Indeed, there may exist two major systems. A general ability to extract patterns from stimuli (PE) and a combinatorial, hierarchical, and recursive system (CHR) for generating complex forms of thought. Daniel Kahneman, a behavioral economist, distinguishes between something he calls *System 1 thinking*, which is intuitive and immediate, from *System 2 thinking*, which is deliberate and logical. However, his model doesn't capture the kind of pattern recognition forms of thought that organic beings are particularly good at (and, more recently, digital computers) as well as the myriad kinds of thought processes employed by humans and other animals. What Professor Kahneman is calling an intuitive and immediate, that is, an elemental cognitive and affective process, may actually be an instinctual response or an *ancestral memory of a highly plastic learned response to the environment* that was sculpted into an instinct by way of evolution. Intuition doesn't tell us anything useful because intuition itself is poorly understood and might actually be quite different from how it is generally defined as "reaching conclusions based on nonconscious processes of reasoning." But, plasticity-first models may help explain the expeditious evolution of behavioral and anatomical changes in higher-order primates. Epigenetic factors underlying behavioral plasticity—changes in the way genes are expressed—regulate protein-coding genes and include DNA methylation (which can change the activity of a segment of DNA), genetic imprinting (which influences genetic alleles from the male or female parent), noncoding RNA (a segment of a RNA molecule that does not code for proteins), and histone modification (methylation of histones that organize and order DNA units) that have significant effects on gene expression. Moreover, just because something is immediate doesn't mean it isn't deliberative because we often don't know, understand or recognize in ourselves or others the history of thinking behind a topic or thing that we have been thinking about, or about to think about, as the end stage of some reasoning process. C. S. Pierce, the American logician, maintained that intuition was actually a mirage because all mental action derives from inference and there is no cognitive stage that precedes all others. Since thinking is always thinking in signs (symbols), according to Pierce, a thought can only be interpreted in what came before it and in what came after it or otherwise it is bereft of any semantic content or meaning. Likewise, *introspection*—observation of one's own mental processes—derives not from a purely internal soliloquy with ourselves, but hypothetical or abductive reasoning about our thoughts and feelings from facts derived from the known external world. Let's take a simple, everyday example of a hypothetical, abductive inference.

> If all the beans in this bag are white, and these beans in my hand are white too, then they probably came from this bag.

Hypothetical reasoning is about some unknown, by definition. If you meet a person at a social event and she or he makes you feel a certain way, your inferences about your internal state when thinking about this person are inferential based on your conscious and nonconscious behavioral and verbal interactions with them. What if, from your interactions with them, you inferred that they liked you? Pierce phrased it this way: "But if A were true, C would be a matter of course. Hence, there is reason to suspect that A is true." In our case, where A = s/he likes me, C = their behavior toward me. We're making hypotheses or abducting conclusions based on actual behavior in the real world. Our knowledge of our own internal states is an inference. If you have ever spent some time on the couch with a psychoanalyst, as I did for 4 years three-days-a-week, you will quickly realize that someone else is making inferences about your internal state of which you, the analysand, may even be unaware. How can that be? You have privileged access to your own private, internal states but your inferences based on your experience about your own mental states may be incorrect or misinformed. Nevertheless, some might argue that psychoanalysis is actually an illusion of the analyst's experience of you, based on their own experience and training, that is, theory and clinical prowess. But if we accept Pierce's stipulations that introspection is nothing more than an inference from external knowledge, than their inferences about you may be plausible—at least, some of time.

Modern hominins, nevertheless, are mentally restricted about what they can think about and what conclusions they can draw—the "absolutely incognizable" to quote Pierce—due to inherent limitations in the MBB. Our cognitive abilities have changed over geological and historical time with the evolution of the brain and body. Inference-drawing is difficult so how do we know that it's an intuition rather than a long drawn out thought process stemming from both unconscious and conscious elements?

> We come to the full possession of our power of drawing inferences the last of all our faculties, for it is not so much a natural gift as a long and difficult art.[2]

In our two systems of thought—pattern extraction (PE) and the combinatorial, hierarchic, and recursive system (CHR)—modern humans and presumably other animals such as infra-human primates as well as fellow travelers such as the brown rat (*Rattus norvegicus*), appear to possess the ability to extract statistical regularities from the environment, that is, make use of a general purpose statistical learning ability as part of the pattern extraction system (PE). This statistical learning ability would have the capability to extract and parse information into meaningful wholes—to predict and generalize from a subset of external stimuli—acquired through the five external senses (vision, audition, gustation, olfaction, and the haptic modality) and probably from at least one internal sense, proprioception (an awareness of sense of motion as well as limb and body position). This ability to extract patterns (termed *pareidolia*; Figure 11.2) would rely on similarity as well as temporal and spatial contiguity. In the figure below, a face is ostensibly extracted from the Martian landscape, by our vision.

Figure 11.2. Pareidolia.
Credit: DancingPhilosopher (https://commons.wikimedia.org/wiki/File:Ajdovska_deklica_-_viewed_from_Erjavčeva_koča.JPG), "Ajdovska deklica – viewed from Erjavčeva koča," https://creativecommons.org/licenses/by-sa/3.0/legalcode.

Statistical Learning

Statistical learning has been demonstrated for speech (words), non-speech sounds, static shapes and shape configurations, tones, and action sequences in both infants and adults. Individuals are even able to track conditional and joint probability for dynamic human action by segmenting action units. The probability of an event given that another event has occurred is called a *conditional probability*. For example, the conditional probability of sneezing if you have an upper respiratory infection is quite high given the symptomatic relationship between the two. On the other hand, *joint probability* represents the frequency that any two elements co-occur and is usually expressed as the probability of event Y occurring at the same time as event X. So, if there is a rock concert (X) there is a high joint probability that young people (Y) will be in attendance. Dynamic actions—opposed to words or shapes—are more difficult to segment because there is much more information to attend to and filter. Nonetheless, humans can segment and pick out these dynamic actions in order to understand *enactive relationships*, such as a curve ball being thrown by a pitcher connecting with a bat swung by a baseball hitter. Individuals spontaneously segment dynamic action into hierarchically structured parts, such as the golf swing I discussed in Chapter 9. What people segment has direct effects on what they tend to remember, what they learn from it, and the intentions behind it. They are not just learning *joint probability*—the co-occurrence of actions and their frequencies—but *conditional probability* or prediction of the relations between dynamic action elements, such as hitting the cue ball sharply towards its bottom edge in order to put backspin on the cue ball so that it returns to its original location after the shot is completed. If you don't play billiards, you might not know this trick. That's statistical learning.

We can therefore say, that there is an associative system, that is, a pattern recognition system (PE), that relies on similarity as well as temporal and spatial contiguity, to extract statistical regularities from our surroundings and may be very useful in unknown and constantly changing environments. In addition, it can use distorted, degraded, and incomplete information to extract these regularities. Something like a representative heuristic, to make use of Daniel Kahneman's ideas again, that relies on representativeness of an event in our surroundings to infer properties of the environment. How representative is the Boho style of dress in the US population? If you live in downtown New York City (Manhattan), it is more common than you might think.

Modern humans, according to Ray Kurzweil, may store a billion patterns in memory—although that number is probably greatly exaggerated as scientific estimates of chessboard patterns memorized by grandmasters in chess are in the thousands—of which about two-thirds comprise common sense knowledge (What does an evergreen tree look like?) and the remaining third consist of more general and extensive professional knowledge. For example, in medical diagnosis, professional knowledge relies on patterns of symptoms to identify a disease or disorder. In law, the fact pattern identified in the initial discovery process helps the attorney construct a legal case. Indeed, artificial intelligence programs are getting better and better at pattern recognition using *deep learning* (a bit of a misnomer) to recognize voices or images; related forms of image analysis to spot particular fingerprints, iris or face identification; harmonic, audio, and movement analysis; biopharmaceutical assay; drug discovery; and the like.

A rule-based system (CHR) on the other hand involves manipulations on relations among symbols. It includes use of general algorithms or rules, heuristics or cognitive strategies, deductive and inductive reasoning, conditional and pragmatic reasoning, categorical and conceptual inference, reasoning by analogy, schemas and scripts as in a restaurant script for obtaining a table and ordering food in an eatery, global and local planning, as well as insight and creative thought. The latter two often rely on partially unconscious processes that may overlap with the pattern recognition system. A rule-based system would also require serial understanding, that is, stringing actions together to achieve a predefined end or praxis.

The interdependence between mind and body, music and speech, image-making and aesthetic and athletic movement, quantitative navigation (number) and sociality share many non-cognitive and motor elements. Rhythm binds together individual nervous systems of humans and other animals. That is, there is entrainment of the CNS, the body, and various bodies together. Although there is some evidence that musical rhythm arose independently of synchronization of utterances during social interaction, they both share the same underlying bases of rhythm and tempo in verbal interaction and in song. Melody, for instance, has both a musical structure—technically referred to as harmony—and an underlying rhythm and tempo. There is also a cultural correspondence between instrumental music and the speech patterns of a culture as well as rhythms and tempos of verbal interaction and their synchronization with collaborating speakers. This is influenced by the physical position of the speakers—or what is called the *deictic relations of speech*—and their social relations with each other such as an individual's power and status and a group's mode of interaction and solidarity. Moreover, music and language share a common neural network that holds temporal patterns and relations as well as pitch in working memory. Musical and linguistic syntax or rules for governing the agglutination of sounds or words over time, are both processed in Broca's area (BA 44/45; Figure 11.3) in the left ventrolateral prefrontal cortex in 95% of individuals.

Music and language, moreover, were present in prehistoric modern humans based on evidence left at paleontological sites. In chimpanzees (*Pans troglodytes*), Broca's and Wernicke's areas are activated when the chimp engages in tool use or what I referred to earlier as *instrumental intelligence*. Instrumental intelligence refers to a subset of bodily-kinesthetic intelligences where a hominid or other animal engages in means-ends relations—that is, uses a means, such as a tool, to gain some end, such as a food item. This has been spoken of as using one object to obtain another or using two objects in unison (a tool such as a stick) to effect an action (obtain grubs from inside a tree stump).

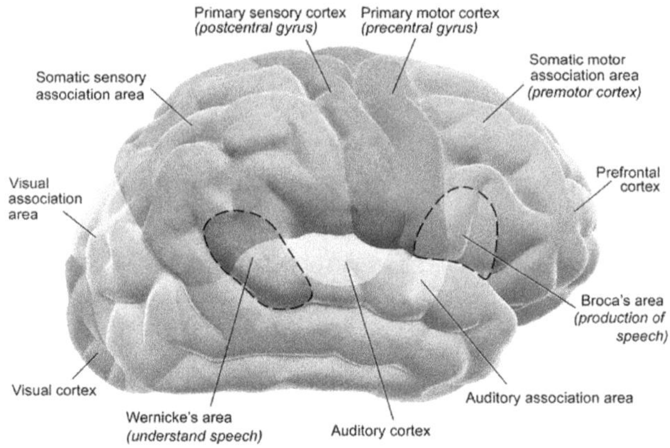

Figure 11.3. Language Network.
Credit: BruceBlaus. (https://commons.wikimedia.org/wiki/File:Blausen_0102_Brain_Motor&-Sensory_(flipped).png), "Blausen 0102 Brain Motor&Sensory (flipped)," https://creativecommons.org/licenses/by/3.0/legalcode.

Sign language and speech, by the same token, both activate Wernicke's and Broca's areas as well as the broader perisylvian area, which is basic to speech and language in modern hominins. Speech is embedded in the motor articulatory apparatus including the tongue, lips, jaw, larynx, and velum (soft palate; Figure 11.4).

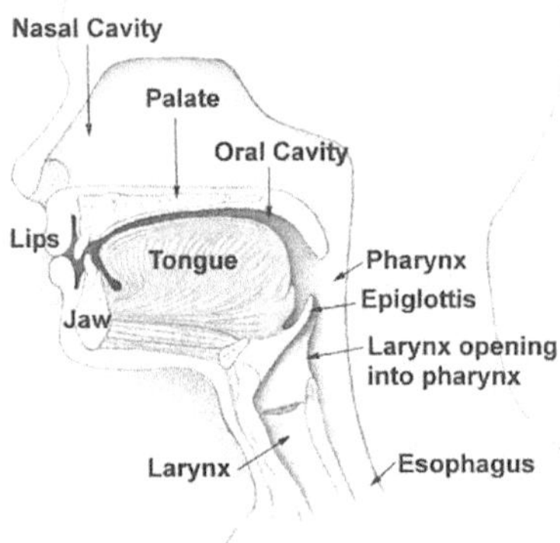

Figure 11.4. Motor Articulatory Apparatus.
Credit: "Head & Neck Overview." National Cancer Institute. Retrieved from https://training.seer.cancer.gov/head-neck/anatomy/overview.html. Public domain.

Moreover, as I presented earlier, motor areas of the brain have cognitive functions. The motor articulatory apparatus, the hands and the body are tools for thinking and communicating with others. Infant-directed speech by adults uses vocal prosody or intonation to facilitate the speaker's intentions and feelings, thus, communicating ideas, concepts, and mood states.

"It's a DOG, Timmy! A DOG."

The sounds of musical notes and phonemes are translated into the abstract motor-neural-articulatory codes of the brain. Mirror neurons, which are activated in several areas in the primate brain when both perceiving a movement or action and performing an action with the same object or body part, are found in the F5 region, that is, the ventral-rostral premotor area (PMVr). The F5 region is found in the ventral or the bottom part of the premotor cortex but rostral to, or in front of, the premotor cortex, which overlaps with Broca's area. The appearance in early hominins of upright posture, of course, freed the hands from locomotion for the manipulation of objects and putatively led to the evolutionary enlargement of the brain in the first punctuated equilibria in hominin evolution (see below). Music, sociality, language, number, and image-making are all co-constituted through movement. Music involves the mind-brain-body (MBB) in tempo, rhythm, and song including birdsong and whale song. Number, at its most elementary, comprises subitization (enumeration without counting), counting, object quantity, number and object equivalence, serial and same/different understanding, relational concepts, as well as the concept of absence. Image-making—visuospatial abilities and their instantiation—is found in infra-human primates and elephants. Sociality, from meerkats to canines to modern humans includes dominance relations and social hierarchies, social relations and roles, social understanding, and emotional and social intelligence as well as its cousin, Machiavellian intelligence. Indeed, sociality and social intelligence emerged long before primates came into existence 65 million year ago. Let's look at the social life of dinosaurs.

Early Sociality in Dinosaurs

About 65 to 145 million years ago during the Cretaceous Period three juvenile dinosaurs huddled together seeking warmth, avoiding predators, and seeking solace in sleep; a behavior in modern animals, particularly birds, known as communal roosting. They comprised a species of early dinosaurs (*Oviraptid*) with long necks and short faces, heads with domed crests, beaks lacking teeth, and locomoted as bipedal walkers. They were about the size of a German Shepard and weighed approximately 45 kilograms or about 99 pounds based on an analysis of their fossils. They may have been siblings from a larger oviparous hatching of 12 or so and were probably around two to five-years-of-age. Two of the dinosaurs were crouched down on their bellies, their arms cradling their heads, and their necks stretched back into their torsos—a pose strikingly similar to ostriches in deep sleep. Presumably, cold weather or even a sandstorm drove them together. But, one thing is fairly certain. They sought out food sustenance in groups, used their crests to attract mates or challenge adversaries, and led a rich social existence long before primates inherited the earth. If social intelligence had anything to do with large brains, it clearly didn't start with early modern humans or even Great Apes.

Early Language

The evolution of language began with the ability to both segment sounds (segmentation) and to compose and combine elements of sound (compositionality and componentiality). Autistic children, for example, may focus on segments of meaning in speech sounds without the ability to combine them into coherent wholes demonstrating a lack of something called *central coherence* in the MBB. The first instances of speech may have initially relied on sound symbolism or *phonosemantics* (words by dint of their particular sound carry meaning, such as an *Acura* car or a *Google* search engine), onomatopoeia (creation of a word from a sound associated with its referent such as "moo, moo"), and vocal imitation. The first words were thus probably heavily influenced by gesture and body language. The emergence in modern hominins of a primitive grammar—an action-object grammar—may have been the result of archaic humans' ability to extract statistical regularities from the environment—a form of pattern recognition; to make predictions from sounds—culminating in a more universal mature grammar of which new kinds of human cognitive abilities and the capacity to exchange more complex information with others evolved. Nonetheless, an action-object grammar and a universal grammar have different evolutionary origins. And the facility of recursion would have vastly extended this range of human communicative capabilities. But language did not evolve directly from movement and gesture alone but from confluence of a number of cognitive systems that predated or evolved coevally with language. Movement, nevertheless, is evolutionarily central to the evolution of all cognitive systems including language.

Dichotomization

An adaptation for the capacity for *dichotomization* arose initially in a distant Great Ape ancestor and enabled the holding of pairs of opposing and dichotomous categories in the MBB. This was the earliest beginnings of the rule-based system for thinking, that is, a combinatorial, hierarchical, and recursive system for generating complex forms of thought in early humans, although remnants of it exist and existed in geological time in other primates and other animals including early oceanic fish. You need something to think with and pairs of opposites (big/small, light/dark, night/day), which the foregoing are natural categories in the world of sentient beings, provide the basic elements of a combinatorial system. Dichotomization could be instantiated in the wet architecture of autochthonous, that is, original or primordial brains, going as far back as 500 million years ago or earlier as this constitutes its basic wiring: "Neurons that fire together, wire together." Some neurons fire together when there's light and others fire together when it's dark. Some fire at large circular objects and some fire at small circular objects, and so on.

Embodied Thought

Manual dexterity and tool use were some of the nascent competencies of *Homo erectus* but appeared to arrive on the evolutionary scene as early as *Australopithecus afarensis*, about 3.4 million years ago, when the first stone tools appeared at paleontological sites. The construction of even simple tools would have required some kind of rule-based system for sorting and combining elements. This was also the first appearance in hominins of *embodied cognition* or *use of the hands and body to think with*, as opposed to just manipulating objects.

Originally, hands and other appendages of the body had evolved as adaptations for manipulating objects, both parts of the body and objects external to the body. But tool use implied that a third object had entered the picture within the triumvirate of hand, object, and tool. Tool use was a byproduct of evolution, not its object. The use of tools were *spandrels*. A spandrel is a specific kind of evolutionary adaptation known as an *exaptation* that wasn't shaped by natural selection but came into being by being *a byproduct of evolution* and was co-opted for its use in the current environment. Tools constituted the beginnings of technical or technological intelligence and a form of bodily-kinesthetic intelligence known as *instrumental intelligence*. Objects to think with. Since then, in human history, the evolution of the brain and the evolution of technology have bootstrapped each other and ratcheted up human mentation. This tripartite relationship, moreover, is analogous to Pierce's distinction among the terms sign, object, and interpretant.

> A sign is something, A, which brings something, B, its interpretant sign, determined or created by it, into the same sort of correspondence … with something, C, its object, as that in which itself stands to C.[3]

In embodied thought, the hand or some other body part enters into a correspondence with an object through the use of a tool that transforms it. It is directly analogous to language—that is, a word, object, and interpretant—where a word is a sign that refers to an object (person, thing or event) and is attached to a meaning (interpretant). This is true for images, diagrams, musical notes, dance steps, concepts and ideas, numbers, and the like. The tool, or use of the tool, was the first hominid interpretant and it furthered more complex human and animal cognitive abilities. However, before signs (symbols) became pervasive in human cognition, earlier hominids and other animals relied on *indices* (Figure 11.5) or some physical connection to an object—such as footprints in the sand—and *icons* (Figure 11.6) or some likeness or semblance to the object. Icons are signs whose outward form expresses their underlying meaning such as early cave paintings.

Figure 11.5. Footprints.
Credit: Hansueli Krapf. (https://commons.wikimedia.org/wiki/File:2012-01-09_15-27-34_Spain_Canarias_Jandía.jpg), "2012-01-09 15-27-34 Spain Canarias Jandía," https://creativecommons.org/licenses/by-sa/3.0/legalcode.

Figure 11.6. Cave Art.
Credit: Prof saxx (https://commons.wikimedia.org/wiki/File:Lascaux_painting.jpg), "Lascaux painting," https://creativecommons.org/licenses/by-sa/3.0/legalcode.

Socialization and social skills had evolved in earlier hominids, most likely in *Homo habilis* or even earlier in Australopithecus with the first elements of the *mentalizing network* coming online. These were primary adaptations in human evolution. The mentalizing network is theorized to be made up of the anterior temporal cortex and the superior temporal sulcus (STS). The STS is the jack-of-all-trades, so to speak.

> In line with anatomical evidence from tracer studies, we propose that the function of the STS varies depending on the nature of network co-activations with different regions in the frontal cortex and medial-temporal lobe. This view is more in keeping with the notion that the same brain region can support different cognitive operations depending on task-dependent network connections, emphasizing the role of network connectivity analysis in neuroimaging.[4]

The mentalizing network is important in understanding the internal states of others as well as their accompanying actions. But, along with the mentalizing network evolved the *amygdala network* for detecting socially relevant stimuli comprising the amygdala, the orbitofrontal cortex (OFC), and the temporal cortex. The *empathy network* emerged for attributing intention to others, which is made up of the insula and cingulate cortices. Mirror neurons and the associated mirror/action network played a role here, too, as they partly facilitated relating a gesture or body posture to an intention.

Modern humans (*H. sapiens*) most likely arose as a geographically isolated speciating population presumably local to Africa (Figure 11.7). Genetic (mitochrondrial DNA) and paleontological evidence suggest modern humans originated in Southeast Kenya and is known as the *African Genesis Hypothesis*. Mitochrondrial DNA or mtDNA is found in the egg cell of the mother (the sperm mtDNA is destroyed during reproduction), is highly conserved in evolution, and has very low rates of mutation. By comparing mtDNA sequences among different species of hominins, molecular biologists construct an evolutionary lineage of past human groups. In the case of modern humans, it has been proposed that they branched off from ancestral *H. erectus* in rapid punctuated equilibrium approximately 200 thousand years ago. As a result of genetic drift—known as the founder effect—modern hominins presumably established a unique ecological niche among early humans. That is, a small group of early humans broke off from the host population with a genetic makeup significantly different, at least in some respects, from the original population. In modern hominins this most likely consisted of a deeper capacity for multiple incarnations of symbolic thought; the appearance of more intricate vocal and gestural language skills; complex movement abilities including the potential for their aesthetic and athletic use; intrapersonal, and social and Machiavellian interpersonal faculties; visuospatial abilities and their dormant capacity for artistic expression; technological skills including both fine- and gross-motor abilities; and numerical and navigational capabilities inclusive of their capability for both approximate and exact mathematical computation. Neoteny, which is a slowing down of development across generations, resulted in an advantageous retardation in development of the CNS in early modern humans and facilitated the development of larger and more modular brains. We mentioned this earlier. A more multi-threaded brain. As *H. sapiens* migrated out of Africa, they replaced *H. erectus* and *H. neandertalensis* and, of the latter, presumably by interbreeding as most present-day Europeans have some Neandertal DNA, myself included, around 3.9% to be exact. You can check out your own genetic makeup for yourself by providing a saliva sample and having your genome analyzed for a small fee (www.23andme.com).

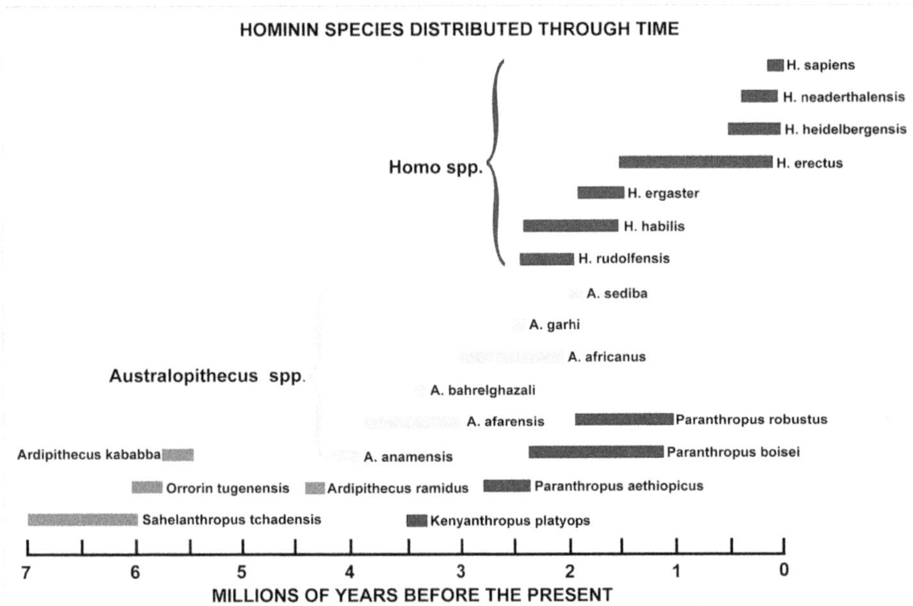

Figure 11.7. Evolutionary Timeline of Hominin.
Credit: Cruithne9 (https://commons.wikimedia.org/wiki/File:Hominin_evolution.jpg), https://creativecommons.org/licenses/by-sa/4.0/legalcode.

The Evolution of Language

The emergence of complex linguistic abilities probably appeared on the scene by 40 thousand years ago or earlier in modern humans. In Chapter 6, we saw how articulate speech in all likelihood first emerged in *H. heidelbergensis* approximately half a million years ago. However, the exact nature of this early speech is unknown although it may have been nearly as complex as that of early modern humans. At this later time, however, there were approximately three groups of hominins inhabiting parts of the world: *H. sapiens*, *H. neandertalensis*, and *H. erectus*, and possibly a subspecies of *H. sapiens*, *H. sapiens denisova*. Vestigial linguistic capacities of these earlier hominins, maybe early protolinguistic abilities, thrust *H. sapiens* into full linguistic production. The perisylvian language network rapidly expanded (Figure 11.8). One hypothesis recently put forth by Robert Berwick and Noam Chomsky at MIT, is that a small genomic change led to completion of a fiber tract or ring connecting the two pairs of dorsal and ventral tracts of the brain in hominins (Figure 11.8). These two (2) pairs of tracts (4 total) coupled the auditory cortex (posterior superior temporal cortex or pSTC or BA22) with the premotor cortex (PMC or BA6, comprising Tract 1, dorsal; BA22 to BA6) to Broca's area for implementing the motor aspects of speech (BA44, comprising Tract 2, dorsal; BA22 to BA44) with retrieval of the words themselves (Wernicke's area, BA22, comprising Tract 3, ventral; BA45 to BA22) to a serial ordering mechanism for syntax (orbital area, BA47) and their meaning in the mental lexicon (temporal pole, BA38, comprising Tract 4, ventral; BA47 to BA38). BA47 has been implicated in the processing of syntax in both oral and sign languages as well as semantic aspects of speech. BA38 has many functions but semantic processing and narrative comprehension are two of them. It brought with it the capacity for *narrative or syntagmatic thought* or what you know as oral storytelling. The prefrontal cortex and other areas of the parietal and cingulate cortices were also involved.

This small genomic change is misleading, however, as it would have required broad and major changes to the cytoarchitecture, connectome, and physiology of the hominin brain over a relatively short geologic time interval if speech miraculously burst on the scene 40–50 thousand years ago, which as it increasingly turns out, is neither consistent with the evolutionary evidence nor with a plethora of neuroscientific studies. There are four tracts as well as

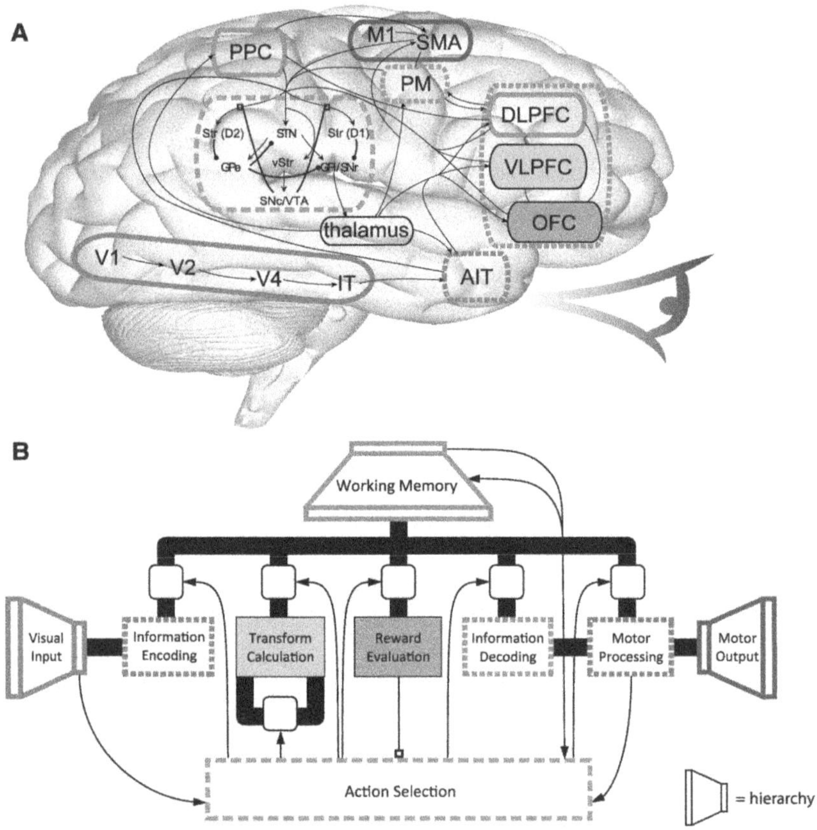

Figure 11.8. Language Areas in the Cortex.
Credit: Public domain.

other major changes that would have been required to parts of the prefrontal cortex and the parietal, temporal, and cingulate cortices that underlie fluent speech. The areas that these tracts interconnected with would have had to undergone significant changes such as phoneme perception in BA22, speech production in BA44, combinatorial ordering of word sounds in BA47, and creation of a brain substrate for word sounds and their associated meanings in BA38, that is, a mental lexicon. Sentence production and narrative comprehension, moreover, would have to be physiologically structured in the brain in areas of the frontal cortex with metacognitive control in the prefrontal areas in BA9 and BA46 in the dorsolateral prefrontal cortex (DLPFC). This highly critical brain area is involved with motor planning and organization, verbal working memory in the left hemisphere, and has important connections to the basal ganglia, specifically the left caudate nucleus for control and monitoring of lexical production. Subcortical connections such as the basal ganglia and cerebellum are also essential for speech production. Word generation, for instance, is bound up with both the left and right lobules of cerebellar region VIIa along with several other adjacent areas in the cerebellum. These areas, including cerebellar regions Crus I/II, L VIIb, VIIIa, are associated with the *default mode network*, which is activated when people are mind-wandering and not engaged in a specific intellective activity, such as *talking to themselves*. Lobules I–IV in the left and right cerebellar hemispheres are involved in auditory working memory and lobules V–VI are implicated in naming, reading, and repetition of word and letter sounds. All are essential for speech production and comprehension. These various structures would have had to be in place and connected to all the major cortical areas for speech to emerge in hominin evolution. So, would an extensive area in the inferior temporal cortex, which is specialized for storage of different kinds of words and concepts such as tools, proper names, and physical objects. Unless it evolved in the last 500 thousand years ago, it wasn't necessarily present when the first words appeared. Indeed, all of these structures are highly interconnected in order to function in producing symbolic speech in multiple cortico-striatal loops. The orbital frontal cortex (OFC), with vital connections to the emotional processing areas in the brain (amygdala, insular cortex), would have had to be in place for proper emotional content of speech to be exhibited in interactions with others, as well as the social nature of the interactions themselves to be co-constituted with a speaker-hearer. On top of this, motor commands must be assembled to control output of the speech-articulatory apparatus, which consists of more than 100 muscles that produce intelligible language. This musculature includes the supralaryngeal structures comprising the vocal tract for articulating words, the abdominal and thoracic muscles for regulation of respiration essential to articulating word sounds, the laryngeal accessory muscles necessary to produce vibration of the vocal folds, along with linguistic and cognitive executive control for the production of meaningful speech and language. This system would have had to evolve over a large swath of geologic time for articulate speech to first appear on the evolutionary scene half a million years ago or so.

At the same time there was coeval emergence of the inferior parietal cortex potentiating the capacity for subitization (enumeration without adding things up), formal counting, and full-blown numeracy. Syntagmatic thought was counterposed with *paradigmatic thought* (category and concept creation), which initially constituted the capacity for seeing patterns in nature or pareidolia, a very useful ability in procuring food and tracking other animals, among other things. This ability for pattern-seeking or *pareidolia* was a precursor for image-making, which the current evidence demonstrates may have appeared around 120 thousand years ago or earlier in very early prehistoric art.

Evolution of Consciousness

We know that perceptual consciousness—primary consciousness or awareness of one's environment—may have arisen in vertebrates as early as 530 million years ago and was probably not a later cataclysmic event as some have suggested. Self-awareness or secondary consciousness may have evolved only in complex vertebrates including canines; cetaceans such as dolphins, porpoises, and whales; chimps and bonobos; elephants; and magpies and possibly other corvids. There are specialized neurons (spindle or *von Economo neurons*) that may contribute to it in the anterior cingulate cortex (ACC), the fronto-insular cortex (FIC), and the DLPFC. These areas, as a whole, are important for attributing intention and sundry mental states to others, known as *theory of mind*, that is presumed to be absent or severely compromised in autistic spectrum disorders, as well as aiding the processing of negative and positive emotions, including empathy towards others. Von Economo neurons are found in dolphins, elephants, macaques, raccoons, and whales and may have similar functions. Interestingly, when von Economo neurons are malfunctioning or destroyed they result in pathological states, which we see in various social-cognitive disorders and diseases such as frontotemporal dementia, schizophrenia, and autism.

Modern humans also have the capacity to think about thinking or engage in complex *metacognition* and that appears to be unique in both the animate and inanimate worlds (digital computers). Secondary consciousness or self-awareness would have enabled inner thought or consciousness of one's own thought processes and the emergence of the brain's default and salience networks. These networks included the *mind-wandering network (default)* I discussed previously, as well as the *network for filtering and detecting relevant stimuli (salience)* made up of the dorsal anterior cingulate cortex and the anterior insula. These networks would have created the capacity for introspection, inferring the actions and intentions of others, as well as myriad unconscious processes underlying complex cognitive and emotional abilities. But extended consciousness as it has been called (self-awareness), would have had the effect of expanding the range of habitable niches and the disparate environments and ecologies that early modern humans could exploit for their own survival advantage.

Confrontational Scavenging, Collaborative Foraging, and Communal Living

What ecological changes in relation to the environment would have driven the genomic and brain transformations in the early hominin brain? One alleged possibility is the effect of *confrontational scavenging* on evolution of the hominin brain. This is a social argument similar to the social grooming theory, the collaborative foraging thesis, and the social brain hypothesis or the belief that living in large groups was the main driver of the increase in size of the primate brain. All of these promulgate the view, in one form or the other, that social interaction and cooperation drove the increase in size of the hominin brain and generally attempt to explain the emergence of virtually every cognitive ability. Human symbolic abilities and reasoning are thus the result of something like the *interiorization of dialogical exchange* between individuals by way of social interaction and cooperation. That is, we incorporate in our minds the thoughts and conversations of others through continuous social communication and that gives rise to the panoply of human symbolic capacities. These arguments are generally applied to language but there are many other complex symbolic abilities that they barely mention, if at all. Does social (manual) grooming found in other primates explain the emergence of music? Does collaborative foraging explain the advent of numeracy? Does confrontational scavenging elucidate the introduction of image-making or aesthetic movement into early hominin culture? Sociality and its components—reading intentions in others, introspecting on one's emotional state, expressing feelings, getting others to go along with your plans—are the stuff of emotional, social, and, it's close cousin, Machiavellian intelligence. While sociality in other animals arose early in vertebrate evolution—like cooperative hunting in wolves and communal roosting in dinosaurs—and is an important aspect of higher-order complex cognitive abilities, by itself it does explain the emergence of these abilities. It doesn't even explain the emergence of language, which is generally the intent of these socially-rooted theories. There are many aspects of social interaction, why should social grooming alone be the central arbiter of speech and language? The contention is that social or manual grooming begat gossip or vocal grooming and that's how language got started. But, the connection from manual grooming, which primates engage in, to social (vocal) grooming or gossip in modern humans, is vastly underdetermined. Was confrontational scavenging the only way that early hominins procured food on the open savannah? They may have also passively scavenged for food or actively sighted and tracked prey such as small animals as well as abundant plant resources. Large prey may have been passively or confrontationally scavenged because of the lack of the ability to take down large prey even with a group of individuals working loosely together. All of these strategies of foraging for food may have been employed by early hominins. But confrontational scavenging wasn't the only way that early humans interacted or sought sustenance. Persistence hunting was probably widely practiced in early hominins because one person or a group of humans could simply wear down prey in the hot African sun and then deliver the fatal blow, an advantage unique to humans with superior bodily cooling ability through sweating. Another method was forest gardening, which probably extends back in time to early Australopithecine cultures. Vine and tree species were introduced along river banks and wet foothills and may have been even later incorporated into simple gardens.

Think of all the ways that humans interact alone and with others and much of that was part and parcel of our ancient ancestor's hunter-and-gatherer-like cultures. For instance, communal living, the division of reproductive labor and cooperative care of the young; physical and sexual interactions with the members of the same species; mutual grooming and interaction; protection from predators, hunting and food gathering; food preparation; fashioning clothing and footwear; securing places to live and constructing livable habitats; exploring new environments; play with oneself and others; preservation of hearth and home; worshiping of spirits and gods, and so on. Indeed, one suggestion is that early hominins groups were animated by the need for emotional bonding among children and caregivers and that females led the charge.

> If females formed the core of these early groups, and language evolved to bond these groups, it naturally follows that early human females were the first to speak. This reinforces the suggestion that language was first used to create a sense of emotional solidarity between allies.[5]

Let's imagine, for a moment, that that we're studying the social interactions of present-day humans 10,000 years into the future. There are many ways that these modern humans interacted, but what if we chose to research their shopping behavior at supermarkets. We might find they traveled in groups (parent and child), conversed about, among other things, their item choices, and used a wheeled cart to place food items of all kinds sourced from all over the world. We would notice in our evolutionary studies that they used a form of monetary exchange to acquire these food items and that they drove off in a wheeled vehicle when they were done. But, could we say of the relatively small fraction of social interactions of modern humans that food-buying comprises, would give us any deep sense of what drove their increasingly large or, more properly, highly organized, distributed, and cross-wired brains? The same could be said of the impact of computers and other technological hardware on present-day modern humans. There are so many cultural reverberations on humans and they are so interrelated that it is difficult to tease out what computers alone may add. The idea is that new behaviors—responses to cultural change or environmental challenges such as the introduction of various kinds of technologies into the human environment—are incorporated into the genome or the genetic makeup of a species. It is often referred to as the *Baldwin Effect* (1896) after the American psychologist, James Mark Baldwin. In contemporary evolutionary developmental biology or *evo-devo*, some trait, such as computer use, is integrated into the organism's behavioral repertoire. That is, some behavior is incorporated into the genome of the organism because it confers some adaptive advantage in the organism's interaction with the present environment.

Even infra-human primates such as chimpanzees and orangutans, however, have highly complex social lives. How can one pick out just one aspect of that lived life and suggest that *that* drove brain evolution and frequently discussed in terms of increasing size? Actually, size as I said before is, in itself, is a poor barometer of mental complexity. Sperm whales, elephants, and dolphins all have larger brains than modern humans and mormyrid electric fishes have a brain-body ratio (called the Encephalization Quotient; EQ not IQ) similar to humans. Evolution in absolute size of the brain is misleading as brains change in internal organization as they evolve in size including fractionating into smaller subunits. That's because as brains increase in size—add more neurons—the density of connections tends to decrease, forcing brains to become more modular. This is just another way of saying that as the brain gets larger, groupings of cells that were once near each other are now farther away. So, it would have been best for those proximate groupings of cells to gather together and specialize for local functions, such as detecting hues of red, and then each uniquely contribute as a collection of separate groups to the overall detection of color. This modularity leads to further specialization of structure and function of the brain (Figure 11.9). Hence, modularity is an adaptation that enhances multitasking in the brain. Multitasking, as it turns out, may be the key after all.

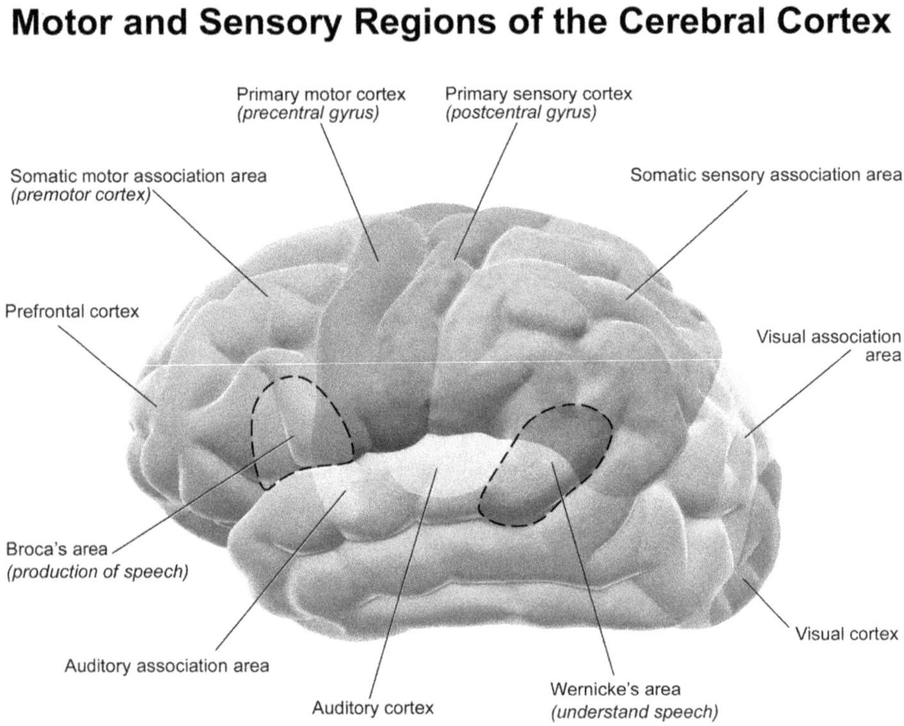

Figure 11.9. Modularity in the Brain.
Credit: BruceBlaus (https://commons.wikimedia.org/wiki/File:Blausen_0102_Brain_Motor%-26Sensory.png), "Blausen 0102 Brain Motor & Sensory," https://creativecommons.org/licenses/by/3.0/legalcode.

Evolution of the Brain and Body

Many aspects of the structure and function of the brain are conserved across species including topological position in relation to other brain regions, connectivity, and histochemistry. Individual brain regions tend to evolve in a predictable manner with absolute brain size. That is, individual brain regions are disposed to evolve in concert with related brain regions or what's called *concerted evolution*, since discrete mental functions are highly distributed in vertebrate brains as a result of modularization. You might think of these as *epigenetic cascades* or a single morphogenetic event that sets off a string of related events such as expansion of the perisylvian region for language.

What is more, the later in development a brain region matures, the larger it is likely to be because of delayed development or *neoteny* as I iterated previously. Ontogeny—individual development of a species, and phylogeny—evolutionary development of a species, are interrelated since disparities in both the rate of growth and developmental timing are the fodder for natural selection. The relative size of a brain region reflects selective adaptation. Take, for example, the lateral prefrontal cortex (LPFC). It's disproportionately larger in modern humans than any other primate species and extensively wired with other parts of the cortical and subcortical parts of the brain. That's because it evolved late in hominid evolution and therefore has very important functions that were selectively adapted. Hence, we saw earlier how the dorsal part of the lateral prefrontal cortex (DLPFC) is essential to speech and language, but it is also important in visuospatial working memory, risky decision-making (cost/benefit analysis in making choices), deductive and syllogistic reasoning, and aspects of social cognition such as inferring intention in others. It has important symbolic functions not just language or social intelligence.

Let's step back and look at some important evolutionary developments in early mammals: Upright posture, endothermy, large brains, and oxidative metabolism. Early mammals included monotremes, marsupials, and placentals; egg-laying, female pouch-carrying, and live birthing mammals, respectively. Cold-blooded ectotherms such as reptiles evolved first with their bodies subject to the vagaries of the outside temperature in the environment. *Endothermy*, however, allowed early mammals to internally generate heat through oxidation or loss of elections. Here's a sketch of the biochemical process.

$C_6H_{12}O_6 + 6\ O_2 = 6\ CO_2 + 6\ H_2O$ + heat (loss of 3 electrons produces heat)
Where:
$C_6H_{12}O_6$ is glucose (extracted from ingested nutrients)
Oxygen (O_2) is the oxidizing agent that catalyzes the chemical reaction creating carbon dioxide (CO_2), water (H_2O), and generating heat.

Oxidative or aerobic metabolism—creation of energy from the breakdown of consumed nutrients into ATP (adenosine triphosphate)—was a significant evolutionary development because it allowed movement and locomotion to operate independently from respiration. One molecule of glucose produces upwards of 30 molecules of ATP in oxidative metabolism, which is significantly more energy (up to 15 times more) than *anaerobic metabolism*—metabolism without oxygen acting as an oxidizing agent—found in simpler vertebrates. It can also be stored for later use, which became an extremely important evolutionary development.

Early mammals also lost their postdentary trough, which formed the middle ear in earlier animals and, as a result, the middle ear bones separated from the lower jaw allowing them to eventually tune-in to high frequency sounds such as music and language. Earlier protomammals relied on bone conduction alone to respond to acoustic stimuli, similar to the deaf percussionist, Evelyn Glennie, I discussed in Chapter 7. The brains of early mammals, however, were small—less than 5 grams—and *lissencephalic* or smooth on the surface, as opposed to complexly folded or *gyrencephalic*, as in modern primate brains. Due to mechanical constraints on living tissue—the physics of the body—as brains become larger the cortex becomes more folded in order to increase the surface area of the underlying tissue allowing for more rapid oxidation and nutrient flow. This is known as the *mechanical constraint hypothesis*. That is, a *radial neocortical column* appears consistently across mammalian species because that particular grouping of neurons in a structural relation with one another provides the *optimal computing unit*. Jeff Hawkins, an engineer whose company created the PalmPilot and the Treo personal digital assistants (PDAs) during the 1990s and early 2000, has been working on a model of neocortical function. In a recent paper, his team at Numenta, an artificial intelligence company, theorized that within the sub-granular layers of each of these neocortical columns, the location of a complex "object" (person, thing or event) being attended to in three dimensions is calculated and combined with sensory information. This information as a whole is then fed back into the cortical network in order to reach a consensus about what the "object" actually is and learn more about the world. He believes that grid-like cells in the hippocampus that inform us about the location of objects in relation to ourselves, so-called *spatial memory*, also operate throughout the neocortex to create actual *concepts* of objects (derived from location and sensory information) suggesting that they are *enactive* and *spatially informed*. Nonetheless, besides mechanical constraints there are many phylogenetic (e.g., descent with modification that results in a common ancestor), ecological, and genetic restrictions and limitations on mammalian evolution.

Energetically the average human brain draws about 16% of the energy of the body—the gut, by comparison, draws about 70%—which is actually quite a bit for a roughly three-pound (1400 grams) object. Resting metabolic rate scales with absolute brain size suggesting the close association between the brain and the body. But as brain areas enlarge, they also become more laminar. *Laminarization* refers to integrated sheets of neurons in the brain that are arranged in layers and, as a result, shorten white matter pathways that carry information to and from grey matter regions, thus decreasing axonal and dendritic wiring and speeding up information processing. The complexity of laminar organization, not size, is one difference between hominin and dolphin brains. Whereas humans have six layers of laminae in the cortical mantle, cetaceans like dolphins, lack layer IV indicating a different connectome (wiring diagram) and cytoarchitecture (cell structure). Similar differences are found in avians and mammals, including hominids. Owls, for instance, have a homologue of the human neocortex but it is not strictly six-layered as in humans, but consists of the *pallium* or dorsal telencephalon (cortex) that is semi-layered (roughly four layers) and laminar. Down the mammalian evolutionary path to modern humans, the *pallium* evolved into the neocortex, hippocampus (encoding memories), amygdala, claustrum (possibly playing a regulatory role in consciousness), and the olfactory cortex.

On the other hand, the dorsal cortex of turtles—*of all species*—forms a missing link between the dorsal pallium of amphibians and the neocortex of mammals, the latter which added new connections, realigned incoming axons, and brought forth new classes of neurons to the pre-existing turtle brain. Where would we be without the hard-shelled turtle? Interestingly, turtles have been around for about 220 million years (that's right, *prototurtles*) and are relatively intelligent. They easily learn to navigate the environment including mazes in the lab, acquire novel kinds of skills, demonstrate long-term memory, and seemingly engage in play. Many are going extinct, unfortunately, because they face many threats in their environments including habitat destruction, harvesting for consumption (yes, many cultures eat turtle meat), and their popularity in the pet trade. On the other hand, there are 328 species of turtles (sea, freshwater, and terrestrial) so they are not going away anytime soon. Turtles are, nonetheless, an important missing link in brain evolution and we owe something to them in our brain anatomy. That's probably why everyone appears to have a natural fondness for them. They're part of us.

Corvid Brain and Intelligence

In many avian species, their *encephalization quotient* or EQ (the ratio of brain mass to body mass) is similar to higher primates in spite of their different brain organization. Calling someone a bird brain may even be a compliment in some cases. Corvids (crows, jays, magpies, and ravens) have uncommonly complex cognitive abilities. Ravens, for instance, anticipate events in the future (up to 17 hours later) including their character, time, and location based on past experiences. They aren't merely engaging in *prospective memory*—remembering to do something in the future—but flexibly planning future behavior, such as what tool they might need or what object with which to barter, and they may imagine themselves doing this at some future time. For instance, in deciding between choice and reward—such as a few seeds now or a grasshopper later today (yes, they love grasshoppers, among other things)—a raven will take temporal factors into account when choosing between different choices and expected rewards. Accordingly, they exercise self-control eschewing impulsive habit. Their complex cognitive abilities evolved to capitalize on the physical requirements of the environment that corvids inhabit. Availability of food sources and when and where to find them as well as how to make use of instrumental intelligence to procure them constitutes a budding physical intelligence. Young ravens, recently weaned from the nest, will examine and pick up just about any object that appears potentially useful. Similarly, ravens, using their observational skills, follow the gaze of conspecifics and other birds, remember the location of the other bird's food caches, and retrieve them at a later time exploiting the raven's social intelligence. Various primates also engage in this kind of *planning behavior* as it appears that the ability to plan may have arisen 320 million years ago (mya) in the last common ancestor of birds and mammals or arose separately in each of these lineages. On the whole, corvids practice self-control, barter for food items with a human experimenter using tokens successfully 77.6% of the time (better than a chimpanzee) as well as exchange "gifts" with humans in comparable situations, achieve their objectives often on their first attempt without prior learning (better than a 4-year-old child), plan ahead and generalize across domains, and choose items based on their usefulness to them at some future time.

> It would suggest that some underlying cognitive functions interact in analogous ways in producing domain-general skills enabling corvids to plan across domains. Evolution can reiterate cognitive architectures that facilitate complex behaviors by way of parallelism. ...[6]

Brain size again is a misleading index of intelligence; it's the complexity of organization and that complexity is an effect of modularization, how neurons are actually wired or connected—the *connectome*—and the underlying cytoarchitecture. In avians, neurons are actually smaller than in mammals so that much more can be compacted into a diminutive space. There is considerable neuronal density at roughly 120 million neurons per gram of brain weight. An asset, no doubt, that was a fortuitous adaptation in evolution. In ravens with a brain mass of 14.5 grams, that would comprise a total of about 1.74 billion neurons (Figure 11.10). Quite a significant amount of brain cells available for cognition and other mental activities. Even so, some amphibians can actually regenerate brain areas although there is purportedly regenerative power even in extant humans as research on the hippocampus has demonstrated. Adult brains in living humans may be able to generate new hippocampal neurons that form the core of memory including the consolidation of long-term memory as well as navigational and spatial memory. Those findings have been recently challenged, but the issue is far from settled among neuroscientists.

Avian	Brain Mass	Forebrain (as percent of cortex)
Raven	14.5 grams	80%
Parrot	9.2 grams	79%
Duck	5.7 grams	65%
Chicken	2.5 grams	53%
Pigeon	2.2 grams	48%

Figure 11.10. Avian Intelligence.
Credit: author.

The brain of diverse species scales allometrically with body size and it is easier to understand mathematically. "Allometric" means that the brain and body each develop at different rates.

$EQ = M_{brain}/0.12 \times M_{body}{}^{0.66}$

Where:

M = mass (in grams)

EQ = encephalization quotient

If we constructed a logarithmic plot of brain mass (in grams) on the y-axis and body mass (in kilograms) along the x-axis, the slope would be approximately ¾ or 0.75. That would mean that for every three units or grams we go up on the y-axis there is a corresponding change of four units or kilograms on the x-axis. The resulting distance above or below this diagonal line is the *encephalization quotient* or purported "intelligence" of the species, the EQ being a proxy for intellect (Figure 11.11). As illustrated in the graph below, modern humans have the largest EQ in the animal kingdom, but a number of species are not far behind.

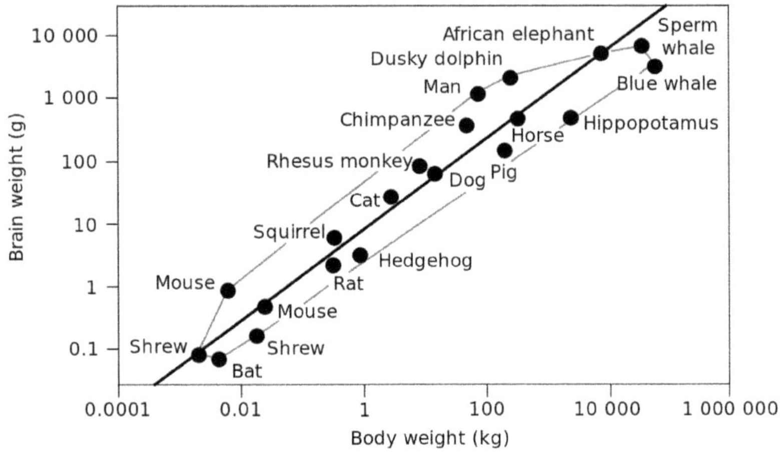

Figure 11.11. Encephalization Quotient.
Credit: User:Benevolent (https://commons.wikimedia.org/wiki/File:Brain-body_mass_ratio_for_some_animals_diagram.svg), https://creativecommons.org/licenses/by-sa/4.0/legalcode.

One reason that modern hominins may have a larger EQ—that is fundamentally different from other animals—is the number of subunits or modules in the modern hominin brain. The *module aggregation hypothesis* states that newer inputs into evolutionarily older parts of the brain form new cortical areas, particularly as it turns out in modern hominin brains that are extensively aggregated to create larger modules. These modules include the perisylvian complex for language (BA 22, 41, 42, 44, 45), the inferotemporal cortex for the storing of proper names as well as names of different kinds of objects such as tools and utensils (BA20, 21, 37, 38), and the inferior parietal lobule comprising the supramarginal gyrus (BA40) and the angular gyrus (BA39) for mathematical operations and bodily and spatial maps. Precision calculation of quantities or numerical intelligence occurs in the left angular gyrus, however, whereas approximation of quantities occurs in the right angular gyrus on the opposite side.

Primate Intelligence

Primate species, 235 of them to be exact, have been around for about 65 million years. They include strepsirrhines or *prosimians* and haplorrhines or *simians* and tarsiers. The prosimian line evolved into its modern variants such as current-day lemurs. Simians evolved much later, around 40 million years ago and include *Old and New World monkeys* (African and South American origins, respectively) such as baboons and spider monkeys; *hominoids* or apes such as gibbons; *hominids* or African apes including chimpanzees, gorillas, and orangutans; and *hominins* or archaic and modern humans (Figure 11.12). Although the DNA sequencing difference in chimpanzees and modern humans is only about 1.2%, that amounts to about 18 million base pairs, which is highly significant. Let's follow the timeline from early primates to early simians to early catarrhines, that is, Simians minus New World monkeys.

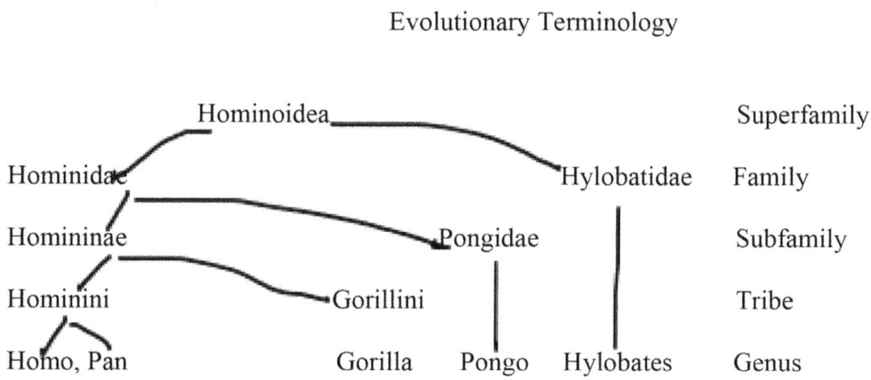

Figure 11.12: Hominoidea Superfamily.
Credit: author.

First of all, a diurnal (daytime) lifestyle became much more common in early primates as well as reduced olfaction as they began to rely more on sight than smell. By the time the Simian *clade* (a group of organisms with a common ancestor) came on the scene there were improvements in daylight vision and, in early catarrhines, complex social behavior had evolved. Absolute brain size increased with the emergence of early primates and relative brain size was magnified two-fold along with larger cerebral cortices to handle all the new information and processing requirements. Metabolic rate surged as did the size of the home range, individual longevity increased, and so did the consumption of fruit or frugivory. These delicious edibles were readily available during certain times of the year and are loaded with antioxidants and various polyphenols that, as it turns out, protect the mind-brain-body (MBB) from disease and disorder. A fortuitous exaptation, no doubt.

Early primates evolved eyes placed in the front of the face and bilateral retinocollicular pathways. There are actually two pathways for vision, not one.

The evolutionarily older one is the bilateral retinocollicular pathway (bilateral because there are pathways to both eyes), which ends in the superior colliculus (optic tectum) in the midbrain rather than by way of the lateral geniculate nucleus to the occipital or striate cortex in the back of the brain, which is called the retinostriate pathway and what we normally think of as vision. The retinocollicular pathway is responsible for control of eye movements and the orienting of visual attention. In essence, it directs the animal to specific objects even in the absence of a functioning visual cortex, a condition known as *blindsight*. In blindsight, studied in humans with damage to the primary areas of the visual cortex (V1), the individual is able to see in one sense, that is, track and grasp objects, but is not aware that he can see because sensory imagery is cut off from conscious awareness. The visual cortex in the back of the brain not only contributes to the recognition of people, objects, and things, but contributes to conscious awareness of what we see. Not only vision, but the primate somatosensory system as a whole evolved with primary, secondary, and tertiary areas that expanded for integration with visual, auditory, haptic, olfactory, and gustatory sensory systems for recognition and discrimination of people, objects, and things. No small feat.

The ventral premotor areas contributed to the increasing use of the hands in order to manipulate objects in much more subtle ways. Visual, tactile (haptic), and proprioceptive feedback from the evolving somatosensory system assisted in developing the use of the hands in primate evolution. Shoulder movement was important in early primates as it began to support and anticipate movement of the hands. As a result, prehensile movements (grasping) were transformed into non-prehensile movements as in a precision grip for manipulating objects. As I discussed in Chapter 2, the thumb became larger with broader fingertip pads and nine independent muscles in the thumb evolved for pronation, supination, flexion, and extension of the thumb musculature. There were changes to wrist and metacarpal bones that supported the use of the thumb and fingers. The right hand—controlled by the opposite hemisphere (left hemisphere) of the brain, so-called *brain and body asymmetry* found in many species—became specialized for throwing and grip strength with the left visual field (controlled by the right hemisphere) contributing sighting and tracking of animate (animal or plant prey) and inanimate objects (tools).

Meanwhile, the orbitofrontal cortex (OFC) in infra-human primates (BA11, BA12, BA13) evolutionarily matured into a reward circuit for external stimuli like sampling the exquisiteness of biting into a fresh piece of fruit. The OFC has extensive connections to many areas in the brain including the amygdala for evaluating external stimuli, somatosensory areas for recording sensory information about the environment, and interconnections with motor areas for control of motor output. The anterior cingulate cortex (ACC) in modern humans (BA24, BA32, BA33) evolved its wiring to register a plethora of information to its owner about the status of her internal state including basic biological drives such as hunger, thirst, and temperature; autonomic functions such as heart rate and blood pressure; as well as contributing to impulse control and various aspects of emotion. The lateral prefrontal cortex (LPFC) became involved in making decisions about one's external environment, particularly social evaluation of conspecifics, as well as possible food sources and other things of importance that one may come in contact.

Archaic Hominin Intelligence

By the time archaic hominins entered the scene 3.5 million years ago there were many complex cognitive abilities already in place. Early Australopithecines were bipedal, it appears that they may have constructed containers for food and water as well as crafted stone tools, probably engaged in some form of primitive speech communication, displayed pronounced imitative abilities, and presumably employed embryonic forms of symbolic thought. On the other hand, control of fire (fire itself occurs in the fossil record 470 million years ago), cooked foods and significant changes in diet and nutrition, as well as the ascendance of more complex kinds of social intelligence, didn't occur until sometime between 1.9 million and 400,000 years ago.

Two bursts in early human evolution have been proposed to account for these changes in early humans. Punctuated equilibrium is ascribed to periods in evolution in which change is fairly rapid compared to much longer periods of relative stasis. Species are always evolving—more or less; some more, some much less—but there are periods in which some aspects of evolution speed up relative to more incremental periods of which the latter are known as lengthy periods of phyletic *gradualism*. Charles Darwin believed that all evolutionary change was incremental but more recent examination of the genetic and fossil record suggests that there are occasionally times when evolutionary change is much more rapid in comparison to the enormous overall slowness of evolutionary time. Mutations, genetic drift, natural selection, recombination of genetic characteristics in sexual reproduction, linkage effects (genes close together on a chromosome ten to be inherited together), and gene flow (migration of genetic characteristics from one population to another) are major factors in evolutionary change.

One burst began about 1.7 million years ago and increased body size in early humans. Australopithecines were actually quite small and relatively short compared to modern humans. The second burst occurred about 400 thousand years ago and resulted in an increase in brain size. There are only so many neurons you can pack into a small brain, there are many physical and mechanical constraints on the brain and body and with increase in size more modularization with all of its benefits began to be expressed in human mentation. One was the purported increase in social skills and understanding, particularly *Machiavellian intelligence*. You see the beginnings of Machiavellian intelligence as increasing population density required heightened competition for resources. As a result, a particular class of social-cognitive abilities emerged including the ability to read others and get them to behave in desired ways often referred to pejoratively as social manipulation. The ability to deceive others through such tactics as lying or withholding information appeared on the scene as did the formation of alliances among groups of individuals to gain the advantage of groups over outsiders as well as the ability to take advantage of others with particular skill sets such as expertise in tool use or greater prowess in hunting skills. The presumption is that these Machiavellian social strategies led to better reproductive and social success and increased the size and complexity of hominin brains. However, you actually see many of these social-cognitive abilities in Great Apes not just in early hominins. We discussed this above in the so-called *social brain hypothesis*.

Another development was the longer period of gestation in early humans—in modern humans, 36 weeks—and the notable increase in brain size after birth. Such a development not only increased the lengthy period of childhood and adolescence compared to earlier hominids, but also provided a unique opportunity for engagement with the environment during this period and the lengthy learning period that ensued with constant input from adults, peers, and the child's culture and environment.

As the human neocortex evolved, it gained direct access to motor neurons that innervated the musculature of the vocal articulatory apparatus including the face, mouth (jaws and tongue), vocal cords, as well as the hands for manual gesture and eventually formal sign languages. These connections between the neocortex and the motor behaviors of the face and hands are known as the *dexterity hypothesis*. That is to say, adroitness in the oral, vocal, and manual modes. Evolution of the corticospinal system in mammals laid the foundation for *cortico-motoneuronal connections* that ratcheted up manual dexterity and, interestingly, are eliminated in development in prenatal mice. The key ingredient appears to be evolutionary emergence of the *PlexA1 gene* (Plexin family of genes) in layer 5 of the motor cortex that underlies cortico-motoneuronal connections in mammals and facilitates fine-motor abilities. Manual dexterity takes advantage of the coupling of corticospinal neurons with motor neurons that undergirds musculature of the hands in primates. This fits in well with the mind-brain-body thesis advanced so far. The interaction of the mind and body makes language, speech, and manual skills possible. Therefore, it has been suggested that human speech would not have evolved without the large lateral prefrontal cortex (LPFC) and its deep neocortical projections to the spinal cord and medulla.

As the primate brain increased in size, much of the modularity of neural networks became serial in design rather than engaging in parallel computation—the neural circuit itself doing many things at once—resulting in a distinctive sequential processing of biological information. Serial circuits have several advantages in that they are more hierarchical than parallel ones and the circuit as a whole operates by processing each output from the prior step ensuring both *feedforward and feedback mechanisms* in information processing and thus allowing for error correction and fine-tuning of the circuit during its use. Here are two linguistic examples of both of these mechanisms.

Feedforward mechanisms are anticipatory in tone, modifying control of a process such as speaking by anticipating what the result should be.

I wanted to capture the struggle it is getting to the top of the mountain, so I decided to use the word "endeavor" rather than "struggle" as I was chatting with my friends.

Feedback mechanisms do just the opposite and they are corrective in nature.

I wanted to say that dating can be laborious, but I think it is really an experience that can give you a better sense of what you want in a mate as I opined to a friend in conversation.

What are some examples of serial circuits? The dorsal and ventral streams in the visual cortex, the thalamocortical circuits that feed into the lateral geniculate nucleus and to the visual cortex, and the dorsothalamical circuits that project to the primary somatosensory cortex (BA5).

The Problem With Ecological Niche-Picking

The need to hunt game in groups and secure food—an example of an *ecological niche*—may have been a vital aspect of confrontational scavenging providing defense from neighboring groups, but the route to language, as I previously suggested, seems convoluted. The same can be said for collaborative foraging and for similar reasons. The need for vocal communication in these collaborative activities is theoretically what facilitated the emergence of speech. According to the confrontational scavenging thesis, a protolanguage first appeared that was made up of sounds that defined something like things and actions. Then, the capacity to communicate about things not present (known in linguistics as *displacement*) as well as future actions. Thus, *prospective memory* of the future purportedly evolved before episodic memory of the past. Or, to put it another way, we first remember to do something in the future in spite of our inability to remember that we might have actually done it in the past. Canines, however, remember complex actions of human adults when either tested at one minute or one hour later. Long before early humans or even Great Apes set foot on Planet Earth, canines had evolved almost 12 million years ago from a common ancestor. *Episodic memory* or something like it, has existed for a very long time and most likely before prospective memory of the future or planning ahead emerged. Dolphins, parrots, canines, and killer whales (*Orcinus orca*) all demonstrate recall of the past. The first actual words, however, were probably *holophrastic* or a single sound communicating both a thing and an action on a thing, as in an individual sound that signified to the group to run from a predator. We see these kinds of communicative intentions in other animals such as meerkats (*Suricata suricatta*, a small diurnal carnivore related to the Mongoose family) who have calls for different kinds of predators. For instance, a terrestrial or an aerial predator. The confrontational scavenging thesis, however, maintains that these first sounds were the first forms of symbolic reference. So, until language emerged, symbolic displacement did not exist. Language was the crucible of thought or, at least, that's the thesis. Then, syntax emerged. And, the capacity for storing all of this somewhere in the brain was apparently already in place.

The collaborative foraging hypothesis gets to more or less the same place by invoking the eminent Russian psychologist, Lev Vygotsky, who, following Karl Marx, argued that thought was the internalization of social dialogue. Once internalized, speech could then be used to guide action. Language comes first circle. These observers of primate behavior and early hominin social interaction argue that human cognition was an adaptation for dealing with social coordination of early hominin groups, but then what was social coordination an adaptation for? Specifically, they argue that three elements were involved in the transition from Great Ape intelligence to the intellective aptitude of modern humans. The first element was cognitive representation or the ability to think with cognitive models of the world and they were roughly in place with the Great Apes. This is a somewhat byzantine way of saying that Great Apes (bonobos, orangutans, chimpanzees, gorillas, and archaic humans) could think with relatively simple concepts such as hot/cold, wet/dry, big/small, and so on. They couldn't communicate it yet with complex words, drawings, diagrams, gestures, and so on, but it was cognitively available to them in some rudimentary form. The second element comprised the capacity for causal inference including proto-understanding of *protoconditional inferences* (if p, then q) and *protonegation* (if not p, then not q); the incipient ability to comprehend *disjunctive syllogisms* (if either p or q is true, and p is false, then q is true); and the capacity for *binary opposition* or what I referred to earlier as *dichotomization*. The first two are elements of simple logic. If it rains (p), the plants will grow (q). If it doesn't rain (p), then the plants won't grow (q). Canines and many other species have understanding of simple causality, A causes B, not just Great Apes. They would also have shared the capacity for dichotomization because that arose much earlier in vertebrate and invertebrate evolution and underlies all of the rest. As regards disjunctive syllogisms, here is an everyday example.

I will choose fresh fruit (p) or I will choose chocolate cake (q).
I will not choose chocolate cake (q).
Therefore, I will choose fresh fruit (p).
Similarly:
It could be P or Q.
It's not P.
Therefore, it's Q.

This kind of embryonic understanding in choice behavior is found in rodents, canids, sea otters, elephants, and probably a considerable number of other species. The *category mistake* is believing that humans and other animals typically make choices based on logical understanding. Notice that making choices instead entails that an animal has a bias in some preference and in mammals this is managed by the frontal lobes. We encountered this earlier when we discussed the *somatic marker hypothesis*. Infra-human primates and other mammals have preferences when choosing just like any other animal. Actually, plants also show preferences or what's called *biological tropisms*, but that's a whole other story.

The third element was the aptitude for behavioral and cognitive self-monitoring, the latter, comprising *executive functions*. Great apes possess these abilities in more or less nascent form although they are found in other species like bottlenose dolphins and canines. Like modern and archaic humans, great apes have a *theory of mind* or the ability to read the minds of others including conspecifics and even modern humans. That is, they understand that the actions of others are driven by beliefs about reality and can infer when these constitute false beliefs. If Chimp A saw Chimp B put a banana in Box 1 but unknown to Chimp B the banana is removed by a human experimenter and placed in Box 2, where does Chimp A think Chimp B will look? Chimp A knows it's in Box 2 and he also knows that Chimp B will look in Box 1 based on a false belief. They literally read this off the bodies of other animals, just as we do, by looking at what the other animal gazes at, their facial expression, and their body language. I discussed the underlying mechanism earlier. It suggests that the evolutionary origins of *theory of mind* emerged with a common ancestor of great apes and demonstrates the mental continuity of all extant great apes including modern humans.

The first step, then, from Great Apes to early hominins as Professor Michael Tomasello of Emory University argues, was facilitated by collaborative foraging, which pushed the cognitive envelope forward with the emergence of cooperative communication and joint intentionality—we are both planning on doing the same thing. These two factors presumably brought on the evolution of the further increase in size and complexity of the early hominin brain. One might question how the need to collaboratively forage arose in the first place (wolves do it, but wolves are highly social to begin with), but the roots of this particular motivation aren't addressed, at least, not directly. I think a more likely possibility is there was cross-talk between the various noetic forms and their instantiation in the early hominid brain or, otherwise, how could *social cognition* have arisen? These various noetic forms must have bootstrapped each other for complex cognitive abilities to evolve. It wasn't born of social interaction and cooperation alone.

The next transition to modern humans, Professor Tomasello suggests, was brought on by the effects of group-mindedness and collective intentionality. Group-mindedness is underdetermined here and only vaguely defined. Does that mean we are all thinking the same thoughts? And, how does that come about? Through common cultural experiences? And, exactly how does that occur? In any event, these social abilities putatively facilitated the ascendance of cultural conventions, group norms, and collective institutions, which are relatively recent. All human thinking was thus essentially derived and constituted from the need for social interaction. Or, did the evolution of complex cognitive abilities in humans drive social interaction, among many other higher-order intellective faculties?

The Collective Behavior of Ants

Let's step back and look at the collective behavior of social insects such as ants. Ants are highly social or eusocial insects of the animal order, *Hymenoptera*, which also includes bees and wasps. They emerged about 99 million years ago with an estimated 22,000 current species and inhabit virtually every environment on Planet Earth. Swarm behavior in eusocial insects has been adapted to robotics research in creating foraging and clustering algorithms for robotic machines as well as network routing in order to solve difficult optimization problems in search algorithms.

Without large brains, but encompassing large numbers of individuals, they engage in relatively sophisticated *collective action* based on the simple behavior of individual ants. There is no management in ant colonies or central control. No individual ant has a plan for the overall collective behavior of the group. How do they then function together? Deborah Gordon, a biologist at Stanford University, has studied red harvester ants for many years in the Arizona desert. Her findings are tantalizingly informative. Each ant's interactions depend on relations with other ants in their immediate vicinity. They communicate with each other through both direct tactile contact—antenna touching—and indirectly through chemical exchanges by way of pheromones. Each individual ant has 14 glands that secrete chemicals known as cuticular hydrocarbons into the environment to communicate with their co-workers or "co-conspirators," if you will. However, their responses to chemical signaling and tactile contact from nearby ants depends on the kind of task they are engaged at that moment including such things as patrolling, refuse cleanup, nest maintenance, defending the colony, and foraging. In computer terminology, we might think of this as the ants' *metadata or contextual code*.

One factor operating is the number and rate of interactions between individual ants. If, for example, food supplies are low, or a new food source is located, more ants will be alerted to switching to foraging behavior. The signaling between ants is not in the individual signal but in the *pattern of signals*. Similarly, the configuration of cuticular hydrocarbons of foragers is different from that of nest maintenance workers channeling individual ants into appropriately different task groupings. If the ant next to you smells right, then, it's your job too although that may change at any moment because there is no caste system in ant colonies because there is no central control. Ant behavior is affected by both the immediate environment of other ants as well as the environment outside the colony such as food sources, interlopers from neighboring colonies, environmental and climate change, and so on. And some of that's registered in the 250,000 neurons of each individual brain of an ant. But, the key to all of this is the ants' *collective intelligence* and their *collective memory*, which radically ratchets up the mental capacity and mnemonic power of any individual ant. Thus, individual ants following simple local rules generate the complex social behavior of large colonies often consisting of millions of individuals. Indeed, we have learned a lot from ant colonies. More than you might think. And, raw social intelligence doesn't explain their group behavior.

Evolutionary Origins of Complex Cognitive Abilities

In the beginning, in the sea, before life in the ocean migrated to land and formed the first animals, unicellular organisms with sensing and efferent properties—the first dual circuits, initially without nuclei and then evolving into eukaryotes, became multicellular and evolved interneurons and complex neural webs to move, navigate, and react to the external world in their aqueous environments. These early creatures of the sea began to parse diversity in the external world into dichotomous categories, which became the beginnings of the first vestiges of mental activity. Features of the external world had now become internal properties of biological systems. It is no doubt consequential that no computer to this day has these properties or could presently acquire them on its own. Nevertheless, motor activity became the basis of mental activity and early intelligence. Dichotomization and movement were the first fundamental features of the mind-brain-body (MBB).

As animals evolved during the Cambrian period 530 million years ago, the first vestiges of sensory consciousness evolved in Animalia with as little as 140 neurons as in the nematode, C. *elegans*. Affective consciousness arose hundreds of million years later around 350 million years ago in early amniotes—tetrapods comprising reptiles, birds, and mammals—in which the first emotional feelings were experienced. Self-awareness or extended consciousness emerged much later in a subset of mammals and primates.

To capture all of this information a sensorium—a sensory manifold—evolved principally around light (electromagnetic radiation), odor (volatilized chemical compounds), sound (vibrations propagating as audible waves of pressure through a medium such as a gas or liquid), touch (a system of neurons that responds to changes on the surface of the body), and taste (substances that react chemically with receptor cells in the oral cavity). But, in the larger organic world as we saw, there are many other kinds of sensory receptors such as biosonar, electroreception, ecto-hormones, and biological compasses.

A second dual circuit arose in evolution to segregate internal sensation from the sensorium and take advantage of the relationship between the two. This had many interesting consequences. Sequencing, organization, and memory of the past became embodied and reflected similar qualities of the external world. One of these qualities was ability to distinguish objects by number and use these attributes to compare quantities as survival often depended on it. Another quality was the emergence of organizational and planning abilities so that animals could anticipate the future like changing environments and fluctuating climates.

As living organisms progressed, pattern-seeking or *pareidolia* became one of the essential ways that organisms gained information and acquired knowledge about physical reality beyond themselves. This became the foundation for *paradigmatic thought and categorization* of the natural world. It also became the core of image-making or visuospatial thinking, which goes back in human history to at least 120 thousand years ago based on recent discoveries of human artifacts. *Narrative or syntagmatic thought*—the conveyance and telling of stories—probably arose first in gesture and then was mapped onto sounds and later became reified in the complex architecture of speech. Rhythm, the bedrock of many bodily and nervous system activities, became linked with sound leading to nonverbal behaviors such as aesthetic movement and music. The first *artistic impulse* arose from movement, embodied in terpsichorean or dance-like behaviors, although sound embodied in music quickly followed.

The underlying cytoarchitecture and connectome evolved for the initial vestiges of creative thought. The ability to relate sensory qualities across different sensory modalities, link an inanimate object to an emotion, transform one movement into another, and associate a sensory quality to an abstract property were creative thought's early scope of influence.

As archaic humans proliferated and passed on and modern humans began to emerge, complex mental processes evolved in the mind-brain-body

(MBB) such as the default and salience networks, the empathy and amygdala circuits, the underlying cerebral architecture for theory of mind, as well as further embodiment of thought itself. This was the dawn of embodied cognition and the deep evolutionary history of thought that first arose in movement. Manual dexterity as well as tool use rapidly evolved. Oral and vocal dexterity lay the foundation for complex syntactical speech. Sociality exploded, although it had a very long evolutionary history dating back at least to the Cretaceous Period or earlier. Yet, it was not because of any specific ecological niche that humans exploited but the effect of many complex environments and interactions among archaic and modern humans, not the least of which was the ratcheting effect of rapidly evolving cognitive and affective streams in evolution. Communal living among early humans and the Great Apes presumably played a much more significant role than any particular ecological niche, such as confrontational scavenging or collaborative foraging, which picked out a small subset of the many ways early humans—male and female, young and old, healthy and disabled, hunter and gatherer—expended their time.

Notes

1. The reference to George Orwell is, of course, from his book, *Animal Farm: A Fairy Story*, 1945. "All animals are equal, but some animals are more equal than others."
2. The quote, "We come to the full possession ..." is from C. S. Pierce, *The fixation of belief*, 1877.
3. The quote, "A sign is something, A ..." is from C. S. Pierce, *Collected Papers*, 1976/1914.
4. The quote, "In line with anatomical evidence from tracer studies ..." is from G. Hein and R. T. Knight, *Superior temporal sulcus—It's my area: or is it?*, 2008.
5. The quote, "If females formed the core of these early groups ..." is from R. Dunbar, *Grooming, Gossip and the Evolution of Language*, 1996.
6. The quote, "It would suggest that some underlying cognitive functions ..." is from A. Nieder, *Inside the corvid brain—probing the physiology of cognition in crows*, 2017.

Bibliography

Allen, C. (1996). Actions and objects: Unequal partners in the evolution of communication. In G. W. Cottrell (Ed.), *Proceedings of the Eighteenth Annual Conference of the Cognitive Science Society* (pp. 215–230). Mahwah, NJ: Lawrence Erlbaum.

Aslin, R. N., Saffran, J. R., & Newport, E. L. (1998). Computation of conditional probability statistics by 8-month-old infants. *Psychological Science*, 9(4), 321–324.

Bellugi, U., & Klima, E. S. (1997). Language, spatial cognition, and the brain. In M. Ito, Y. Miyashita, & E. Rolls (Eds.), *Cognition, computation, and consciousness* (pp. 177–189). Oxford: Oxford University Press.

Berwick, R. C., & Chomsky, N. (2015). *Why only us: Language and evolution*. Cambridge, MA: MIT Press.

Boeckle, M., & Clayton, N. S. (2017). A raven's memories are for the future: Raven can plan for expected future events based on past experiences. *Science*, 357(6347), 126–127.

Botha, R., & Everaert, M. (2013). (Eds.). *The evolutionary emergence of language: Evidence and inference*. Oxford: Oxford University Press.

Bruner, J. (1983). *Child's talk: Learning to use language*. New York, NY: W. W. Norton.

Corballis, M. (1999). The gestural origins of language. *American Scientist*, 87(2), 138–145.

Corballis, M. (2002). *From hand to mouth: The origins of language*. Princeton, NJ: Princeton University Press.

Corballis, M. (2011). *The recursive mind: The origins of human language, thought, and civilization*. Princeton, NJ: Princeton University Press.

Corballis, M. (2017). *The truth about language: What it is and where it came from*. Chicago, IL: The University of Chicago Press.

Dediu, D., & Levinson, S. C. (2013). On the antiquity of language: The reinterpretation of Neandertal linguistic capacities and its consequences. *Frontiers in Psychology—Language Sciences*, 4(397), 12.

de Waal, F. B. M. (2016). Apes know what others believe. *Science*, 354(6308), 39–40.

Diamond, J. (1997). *Guns, germs, and steel: The fate of human societies*. NY: W. W. Norton.

Diamond, J. (2011). *Collapse: How societies choose to fail or succeed* (2nd ed.). NY: Penguin Books.

Diamond, J. (2012). *The world until yesterday: What can we learn from traditional societies?* NY: Penguin Books.

Dick, A. S., Bernal, B., & Tremblay, P. (2014). The language connectome: New pathways, new concepts. *The Neuroscientist*, 2014, Vol. 20(5), 453–467.

Dunbar, R. I. M. (1996). *Grooming, gossip and the evolution of language*. London: Faber and Faber.

Erdal, D., & Whiten, A. (1996). Egalitarianism and Machiavellian intelligence in human evolution. In P. Mellars & K. Gibson (Eds.), *Modelling the early human mind* (pp.139–150). Cambridge, England: McDonald Institute for Archaeological Research.

Fox, K. C. R., Muthukrishna, M., & Shultz, S. (2017). The social and cultural roots of whale and dolphin brains. *Nature Ecology and Evolution*, *1*(11), 1699–1705.

Fugazza, C., Pogany, A., & Miklosi, A. (2016, November 23). Recall of other's actions after incidental encoding reveals episodic-like memory in dogs. *Current Biology*, *26*(23), 32093213.

Gardner, H. (1993). *Frames of mind: The theory of multiple intelligences* (Rev. ed.). New York, NY: Basic Books.

Gazzaniga, M. S., Ivry, R. B., & Mangun, G. R. (2014). *Cognitive neuroscience: The biology of mind* (4th ed.). NY: W. W. Norton.

Gilarranz, L. J., Rayfield, B., Liñán-Cembrano, G., Bascompte, J., Gonzalez, A. (2017). Effects of network modularity on the spread of perturbation impact in experimental metapopulations. *Science*, 357(6347), 199–201.

Gordon, D. M. (1999). *Ants at work: How an insect society is organized.* NY: W. W. Norton.

Gould, S. J. (2002). *The structure of evolutionary theory*. Cambridge MA: Belknap Press of Harvard University Press.

Gregory, R. C. (1998). (Ed.). *The Oxford companion to the mind.* Oxford, New York: Oxford University Press.

Gu, Z., Kalambogias, J., Yoshioka, S., Han W., Li, Z., Kawasawa, Y. I., Pochareddy, S., … Yoshida. Y. (2017). Control of species-dependent cortico-motoneuronal connections underlying manual dexterity. *Science*, 357(6349), 400–404.

Hallinan, E. V., & Kuhlmeier,. V. A. (2008). Ontogeny, phylogeny, and the relational reinterpretation hypothesis. *Behavioral and Brain Sciences*, *31*(2), 138–139.

Hauser M., Chomsky N., Fitch, W. T. (2002). The language faculty: What is it, who has it, and how did it evolve? *Science*, *298*(5598), 1569–1579.

Hawkins, J., Lewis, M., Klukas, M., Purdy, S., & Ahmad, S. (2018, October 13). A framework for intelligence and cortical function based on grid cells in the neocortex. doi: http://dx.doi.org/10.1101/442418.

Herculano-Houzel, S. (2009, November 9). The human brain in numbers: A linearly scaled-up primate brain. *Frontiers in Human Neuroscience*. [online] https://doi.org/10.3389/neuro.09.031.2009.

Hoffmann, D. L., Standish, C. D., García-Diez, M., Pettitt, P. B., Milton, J. A., Zilhão, J., … Pike, W. G. (2018). U-Th dating of carbonate crusts reveals Neandertal origin of Iberian cave art. *Science*, 359(6378), 912–915.

Jackson, M. (1983). Thinking through the body: An essay on understanding metaphor. *Social Analysis*, *14*, 127–149.

James, W. (1890). *The principles of psychology* (2 vols.). New York, NY: Dover Publications.

Johnson, M. (1987). *The body in the mind: The bodily basis of meaning, imagination, and reason.* Chicago, IL: The University of Chicago Press.

Kabadayi, C., & Osvath, M. (2017). Ravens parallel great apes in flexible planning for tool-use and bartering. *Science*, 357(6347), 202–204.

Kahneman, D. (2013). *Thinking, fast and slow*. New York, NY: Farrar, Straus and Giroux.

Kahneman, D., Slovic, P., & Tversky, A. (1982). *Judgment under uncertainty: Heuristics and biases*. New York, NY: Cambridge University Press.

Kimura, D. (1993). *Neuromotor mechanisms in human communication*. New York, NY: Oxford University Press.

Klein, R. G. (2002). *The dawn of human culture*. New York, NY: Wiley & Sons.

Koch, C., & Laurent, G. (1999). Complexity and the nervous system. *Science*, *284*(5411), 96–98.

Kolb, B., & Whishaw, I. Q. (2015). *Fundamentals of human neuropsychology* (7th ed.). Madison, NY: Worth Publishers.

Krupenye, C., Kano, F., Hirata, S., Call, J., & Tomasello, M. (2016). Great apes anticipate that other individuals will act according to false beliefs. *Science*, 354(6308), 110–113.

Kurzweil, R. (2012). *How to create a mind: The secret of human thought revealed.* New York, NY: Viking.

Lopez, O. (1999, June). *The relationship between nonverbal behavior and thought.* Senior Honors Thesis, NIH/Minority Access to Research Careers (MARC) Program. New York, NY: York College/City University of New York, Department of Political Science and Psychology.

Lopez, O., & Seitz, J. A. (1998, November). *The bodily basis of thought: Observational data.* Paper presented at the Annual Meeting of the National Minority Research Symposium, New York, NY

Lyle, K. B., Dombroski, B. A., Faul, L., Hopkins, R. F., Naaz, F., Switala, A. E., & Depue, B. E. (2017). Bimanual coordination positively predicts episodic memory: A combined behavioral and MRI investigation. *Brain and Cognition, 118,* 71–79.

Mansouri, F. A., Koechlin, E., Rosa, M. G. P., Buckley, M. J. (2017). Managing competing goals—A key role for the frontopolar cortex. *Nature Reviews Neuroscience, 18*(11), 645–657.

Meyer, M., & Baldwin, D. (2011). Statistical learning of action: The role of conditional probability. *Learning & Behavior, 39*(4), 383–398.

Mithen, S. (1996). *The prehistory of the mind: A search for the origins of art, religion and science.* London: Thames and Hudson.

Mithen, S. (2006). *The singing neanderthals: The origins of music, language, mind, and body.* Cambridge, MA: Harvard University Press.

Morgan, J. E., & Ricker, J. H. (2008). (Eds.). *Textbook of clinical neuropsychology.* New York, NY: Taylor & Francis.

Morell, V. (2018, February 16). Think "birdbrain" is an insult? Think again. *National Geographic* [online: https://www.nationalgeographic.com/magazine/].

Mountcastle, V. B. (1978). An organizing principle for cerebral function: The unit module and the distributed system. In G. M. Edelman and V. B. Mountcastle (Eds.), *The mindful brain: Cortical organization and the group-selective theory of higher brain function.* Cambridge, MA: MIT Press.

Muto, V., Jaspar, M., Meyer, C., Kussé, C., Chellappa, S. L., Degueldre, C., … Maquet, P. (2016). Local modulation of human brain responses by circadian rhythmicity and sleep debt. *Science, 353*(6300), 687–690.

Nieder, A. (2017, August). Inside the corvid brain—probing the physiology of cognition in crows. *Current Opinion in Behavioral Sciences, 16,* 8–14.

Orwell, G. (1956). *Animal farm: A fairy story.* New York, NY: Signet (originally published 1945).

Paredes, M. F., David, J., Gil-Perotin, S., Kim, H., Kotter, J. A. Ng, C., … Alvarez-Buylla, A. (2016). Extensive migration of young neurons into the infant human frontal lobe. *Science, 354*(6308), 81–84.

Pepperberg, I. (2008). *Alex and me.* NY: HarperCollins.

Pierce, C. S. (1877, November). The fixation of belief. *Popular Science Monthly,* 12, 1–15.

Pierce. C. S. (1976). *Collected papers.* New York, NY: Dover Publications (Originally published 1914).

Ridley, M. (2015). *The evolution of everything: How new ideas emerge.* New York, NY: HarperCollins.

Robinson, G. E., & Barron, A. B. (2017). Epigenetics and the evolution of instincts. *Science, 356*(6333), 26–27.

Roth, G., & Dicke, U. (2005). Evolution of the brain and intelligence, *Trends in the Cognitive Sciences*, 9(5), 250–257.

Rutz, C. Bluff, L. A., Reed, N., Troscianko, J., Newton, J., Inger, R., … Bearhop, S. (2010). The ecological significance of tool use in New Caldonian crows. *Science, 329*(5998), 1523–1526.

Saffran, J. R., Johnson, E. K., Aslin, R. N., & Newport, E. L. (1999). Statistical learning of tone sequences by human infants and adults. *Cognition, 70*(1), 27–52.

Sales-Pardo, M. (2017). The importance of being modular. *Science, 357*(6347), 128–129.

Sanides, F. (1970). Functional architecture of motor and sensory cortices in primates in the light of a new concept of neocortex evolution. In C. R. Noback & W. Montagna (Eds.), *The primate brain: Advances in primatology*. (Vol. I, pp. 137–208). New York, NY: Appleton-Century-Crofts.

Seitz, J. A. (1992). [Review of Emotion Concepts]. *Metaphor and Symbolic Activity*, 7(2), 99–101.

Seitz, J. A. (1994, June). *Thinking kinesically: Theory and practice*. Paper presented at the 24th Annual Symposium of the Jean Piaget Society, Chicago, IL.

Seitz, J. A. (1999, June 4). Selfish sentinels in cooperative mammals [Letter to the editor: "Darwinian selfishness." Response to T. H. Clutton-Brock]. *Science, 284*(5420), 1640–44.

Seitz, J. A. (1999, November 20). Einstein's brain [Letter to the editor]. *The Lancet*, 354(9192), 1822–1823 (with commentary by A. M. Galaburda, Harvard; Jay A. Seitz, CUNY; Robert Salvatori, Johns Hopkins; and reply by the authors, S. F. Wittelson, D. L. Kigar, and T. Harvey).

Seitz, J. A. (2005). The neural, evolutionary, developmental, and bodily basis of metaphor. *New Ideas in Psychology: An International Journal of Innovative Theory in Psychology, 23*(2), 74–95.

Shermer, M. (2011). *The believing brain: From ghosts and gods to politics and conspiracies—How we construct beliefs and reinforce them as truths*. New York, NY: St. Martin's Griffin.

Slon, V., Hopfe, C., Weiß, C. L., Mafessoni, F., de la Rasilla, M., Carles Lalueza-Fox, C., … Meyer, M. (2017). Neandertal and Denisovan DNA from Pleistocene sediments. *Science, 356*(6338), 605–608.

Snowdon, C. T. (2017). Learning from monkey "talk": Nonhuman primates are capable of several key elements of language. *Science, 353*(6330), 1120–1122.

Sorrells, S. F., Paredes, M. F., Cebrian-Silla, A., Sandoval, K., Qi, D., Kelley, K. W., … Alvarez-Buylla, A. (2018, March). Human hippocampal neurogenesis drops sharply in children to undetectable levels in adults. *Nature*, 124, 319.

Stevens, C. F. (1989). How cortical interconnectedness varies with network size. *Neural Computation*, 1(4), 473–479.

Stiles, J. (2008). *The fundamentals of brain development: Integrating nature and nurture*. Cambridge, MA: Harvard University Press.

Striedter, G. F. (2005). *Principles of brain evolution*. Sunderland, MA: Sinauer Associates.

Stuart, S. A. J. (2010). Conscious machines: memory, melody and muscular imagination. *Phenomenology and the Cognitive Sciences*, 9(1), 37–51.

Studdert-Kennedy, M., Terrace, H., & Tattersall, I. (2016, September 29). The birth of language. *New York Review of Books*, pp. 39–40.

Taylor, A. H., Elliffe, D., Hunt, G. R., & Gray, R. D. (2010). Complex cognition and behavioural innovation in New Caledonian crows. *Proceedings of the Royal Society of London* (B), 277(1694), 2637–2643.

Tomasello, M. (2000). *The cultural origins of human cognition*. Cambridge, MA: Harvard University Press.

Tomasello, M. (2014). *A natural history of human thinking*. Cambridge, MA: Harvard University Press.

Trembley, P., & Dick, A. S. (2016). Broca and Wernicke are dead; or moving past the classic model of language neurobiology. *Brain & Language*, 162, 60–71.

Van Essen, D. (1997). A tension-based theory of morphogenesis and compact wiring in the central nervous system. *Nature*, 385(6614), 313–318.

Wade, L. (2017). Claim of very early humans in Americas shocks researchers. *Science*, 356(6336), 361.

Watson, T. (2017, August 31). Roosting dinosaurs are a fossil first. *Nature*, 548, 510.

Yamamoto, H. et al. (2018). Impact of modular organization on dynamical richness in cortical networks. *Science Advances*, 4(11), 1–11.

Zacks, J. M., & Swallow, K. H. (2007). Event segmentation. *Current Directions in Psychological Science*, 16(2), 80–84.

www.ingramcontent.com/pod-product-compliance
Ingram Content Group UK Ltd.
Pitfield, Milton Keynes, MK11 3LW, UK
UKHW021836100926
8556IPUK00004B/384

9 781433 166143